PEARSON

Math
Makes Sense

7

Marc Garneau

John Pusic

Kanwal Neel

Sharon Jeroski

Susan Ludwig

Robert Sidley

Ralph Mason

Trevor Brown

With Contributions from

Cynthia Pratt Nicolson

Margaret Sinclair

Antonietta Lenjosek

Michael Davis

Elizabeth A. Wood

Daryl M.J. Chichak

Jason Johnston

Steve Thomas

Don Jones

Ken Harper

Mary Doucette

Bryn Keyes

Ralph Connelly

Nora Alexander

PEARSON
Education
Canada

Publisher
Claire Burnett

Publishing Team
Enid Haley
Lesley Haynes
Alison Rieger
Ioana Gagea
Lynne Gulliver
Stephanie Cox
Cheri Westra
Judy Wilson

Design
Word & Image Design Studio Inc.

Math Team Leader
Diane Wyman

Product Manager
Kathleen Crosbie

The information and activities presented in this book have been carefully edited and reviewed. However, the publisher shall not be liable for any damages resulting, in whole or in part, from the reader's use of this material.

Brand names that appear in photographs of products in this textbook are intended to provide students with a sense of the real-world applications of mathematics and are in no way intended to endorse specific products.

The publisher wishes to thank the staff and students of Herchmer Elementary School for their assistance with photography.

ISBN-13 978-0-321-43155-4
ISBN-10   0-321-43155-3

Printed in the United States

16 16

Consultants and Advisers

Consultants

Craig Featherstone
Mignonne Wood
Trevor Brown

Assessment Consultant
Sharon Jeroski

Elementary Mathematics Adviser
John A. Van de Walle

Advisers

Pearson Education thanks its Advisers, who helped shape the vision for *Pearson Math Makes Sense* through discussions and reviews of manuscript.

Joanne Adomeit
Bob Belcher
Bob Berglind
Auriana Burns
Steve Cairns
Edward Doolittle
Brenda Foster
Marc Garneau
Angie Harding
Florence Glanfield

Jodi Mackie
Ralph Mason
Christine Ottawa
Gretha Pallen
Shannon Sharp
Cheryl Shields
Gay Sul
Chris Van Bergeyk
Denise Vuignier

Reviewers

Field Test Teachers

Pearson Education would like to thank the teachers and students who field-tested *Pearson Math Makes Sense 7* prior to publication. Their feedback and constructive recommendations have helped us to develop a quality mathematics program.

Aboriginal Content Reviewers

Glenda Bristow, First Nations, Métis and Inuit Coordinator, St. Paul Education Regional Div. No. 1
Edward Doolittle, Assistant Professor of Mathematics, University of Regina
Patrick Loyer, Consultant, Aboriginal Education, Calgary Catholic

Grade 7 Reviewers

Angie Balkwill
Regina Public School Board, SK

Betty Barabash
Edmonton Catholic School District, AB

Lorraine M. Baron
School District 23 (Central Okanagan), BC

Warren Brownell
Moose Jaw School Division 1, SK

Laura Corsi
Edmonton Catholic School District, AB

Tricia L. Erlendson
Regina Separate School Division, SK

Kira Fladager
Regina Public School Board, SK

Daniel Gallays
Greater Saskatoon Catholic Schools, SK

Lise Hantelmann
School District 91 (Nechako Lakes), BC

Tammy L. Hartmann
Simon Fraser University, BC

Jacinthe Hodgson
Regina Public School Board, SK

Mary-Elizabeth Kaiser
Calgary Board of Education, AB

Geri Lorway
Consultant, AB

Rob Marshall
School District 22 (Vernon), BC

Sandra Maurer
Livingstone Range School Division No. 68, AB

Stephanie Miller
School District 41 (Burnaby), BC

Kanwal Neel
School District 38 (Richmond), BC

Jackie Ratkovic
Consultant, AB

Suzanne Vance
Moose Jaw School Division 1, SK

Randi-Lee Weninger
Greater Saskatoon Catholic School Division, SK

Michele Wiebe
School District 60 (Peace River North), BC

Table of Contents

Investigation: Making a Booklet   2

UNIT 1   Patterns and Relations

Launch   4

1.1 Patterns in Division   6
1.2 More Patterns in Division   10
Reading and Writing in Math: Writing to Explain Your Thinking   14
1.3 Algebraic Expressions   16
1.4 Relationships in Patterns   20
1.5 Patterns and Relationships in Tables   25
Mid-Unit Review   29
1.6 Graphing Relations   30
1.7 Reading and Writing Equations   35
1.8 Solving Equations Using Algebra Tiles   38
Unit Review   43
Practice Test   47
Unit Problem: Fund Raising   48

UNIT 2   Integers

Launch   50
2.1 Representing Integers   52
2.2 Adding Integers with Tiles   56
2.3 Adding Integers on a Number Line   60
Mid-Unit Review   65
2.4 Subtracting Integers with Tiles   66
2.5 Subtracting Integers on a Number Line   71
Reading and Writing in Math: Writing to Reflect on Your Understanding   76
Unit Review   78
Practice Test   81
Unit Problem: What Time Is It?   82

UNIT 3 — Fractions, Decimals, and Percents

Launch	84
3.1 Fractions to Decimals	86
3.2 Comparing and Ordering Fractions and Decimals	91
3.3 Adding and Subtracting Decimals	96
3.4 Multiplying Decimals	100
3.5 Dividing Decimals	104
3.6 Order of Operations with Decimals	108
Mid-Unit Review	110
3.7 Relating Fractions, Decimals, and Percents	111
3.8 Solving Percent Problems	114
The World of Work: Sports Trainer	117
Reading and Writing in Math: Writing Instructions	118
Unit Review	120
Practice Test	123
Unit Problem: Shopping with Coupons	124
Cumulative Review Units 1-3	126

UNIT 4 — Circles and Area

Launch	128
4.1 Investigating Circles	130
4.2 Circumference of a Circle	133
Mid-Unit Review	138
4.3 Area of a Parallelogram	139
4.4 Area of a Triangle	143
4.5 Area of a Circle	148
Game: Packing Circles	153
Reading and Writing in Math: Reading for Accuracy—Checking Your Work	154
4.6 Interpreting Circle Graphs	156
4.7 Drawing Circle Graphs	161
Technology: Using a Spreadsheet to Create Circle Graphs	165
Unit Review	167
Practice Test	171
Unit Problem: Designing a Water Park	172
Investigation: Digital Roots	174

UNIT 5 — Operations with Fractions

Launch	176
5.1 Using Models to Add Fractions	178
5.2 Using Other Models to Add Fractions	181
5.3 Using Symbols to Add Fractions	186
Mid-Unit Review	190
5.4 Using Models to Subtract Fractions	191
5.5 Using Symbols to Subtract Fractions	195
5.6 Adding with Mixed Numbers	199
5.7 Subtracting with Mixed Numbers	204
The World of Work: Advertising Sales Representative	209
Reading and Writing in Math: Writing a Complete Solution	210
Unit Review	212
Practice Test	215
Unit Problem: Publishing a Book	216

UNIT 6 — Equations

Launch	218
6.1 Solving Equations	220
6.2 Using a Model to Solve Equations	226
6.3 Solving Equations Involving Integers	231
Mid-Unit Review	236
6.4 Solving Equations Using Algebra	237
6.5 Using Different Methods to Solve Equations	240
Game: Equation Baseball	245
Reading and Writing in Math: Decoding Word Problems	246
Unit Review	248
Practice Test	251
Unit Problem: Choosing a Digital Music Club	252
Cumulative Review Units 1-6	254

UNIT 7 — Data Analysis

Launch	256
7.1 Mean and Mode	258
7.2 Median and Range	262
7.3 The Effects of Outliers on Average	267
7.4 Applications of Averages	271
Technology: Using Spreadsheets to Investigate Averages	276
Mid-Unit Review	278
7.5 Different Ways to Express Probability	279
7.6 Tree Diagrams	284
Game: All the Sticks	289
Reading and Writing in Math: Using a Frayer Model	290
Unit Review	292
Practice Test	295
Unit Problem: Board Games	296

UNIT 8 — Geometry

Launch	298
8.1 Parallel Lines	300
8.2 Perpendicular Lines	303
8.3 Constructing Perpendicular Bisectors	306
8.4 Constructing Angle Bisectors	310
Mid-Unit Review	314
8.5 Graphing on a Coordinate Grid	315
8.6 Graphing Translations and Reflections	320
8.7 Graphing Rotations	325
Technology: Using a Computer to Transform Shapes	330
Reading and Writing in Math: Making a Study Card	332
Unit Review	334
Practice Test	337
Unit Problem: Design the Cover	338
Investigation: Integer Probability	340
Cumulative Review Units 1-8	342
Answers	346
Illustrated Glossary	376
Index	382
Acknowledgments	385

Welcome to
Pearson Math Makes Sense 7

Math helps you understand your world.

This book will help you improve your problem-solving skills and show you how you can use your math now, and in your future career.

The opening pages of **each unit** are designed to help you prepare for success.

Find out **What You'll Learn** and **Why It's Important**. Check the list of **Key Words**.

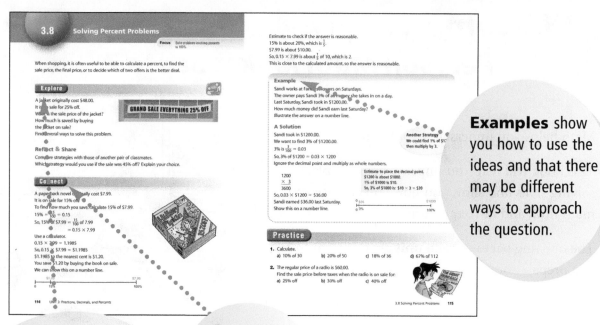

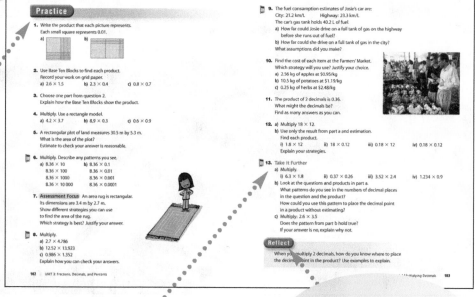

Examples show you how to use the ideas and that there may be different ways to approach the question.

Explore an idea or problem, usually with a partner, and often using materials.

Connect summarizes the math.

Practice questions reinforce the math.

Take It Further questions offer enrichment and extension.

Reflect on the big ideas of the lesson. Think about your learning style and strategies.

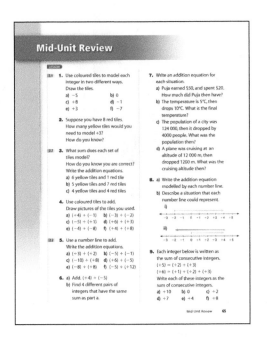

Use the **Mid-Unit Review** to refresh your memory of key concepts.

Reading and Writing in Math helps you understand how reading and writing about math differs from other language skills you use. It may suggest ways to help you study.

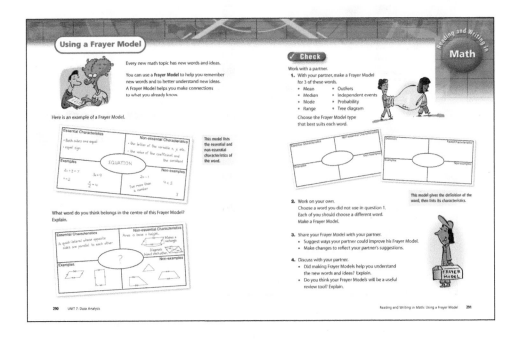

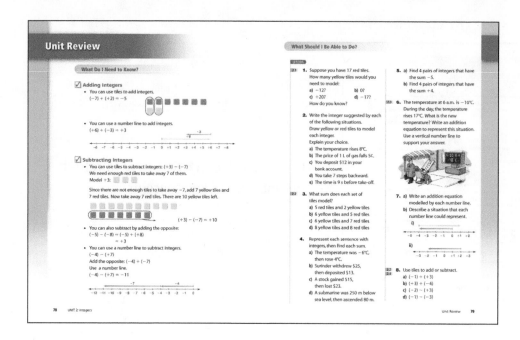

What Do I Need to Know? summarizes key ideas from the unit.

What Should I Be Able to Do? allows you to find out if you are ready to move on. The Practice and Homework book provides additional support.

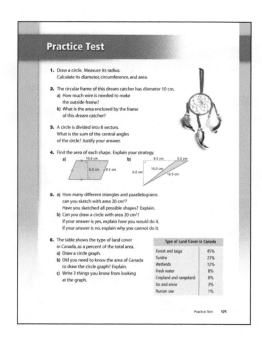

The **Practice Test** models the kind of test your teacher might give.

The **Unit Problem** presents problems to solve, or a project to do, using the math of the unit.

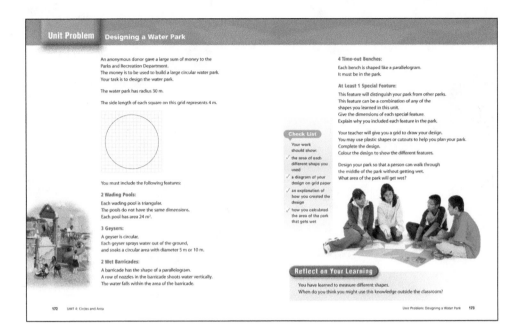

Keep your skills sharp with **Cumulative Review.**

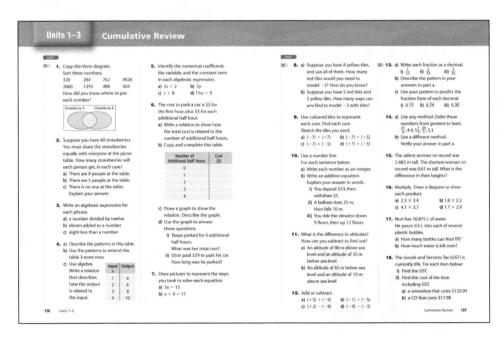

Explore some interesting math when you do the **Investigations**.

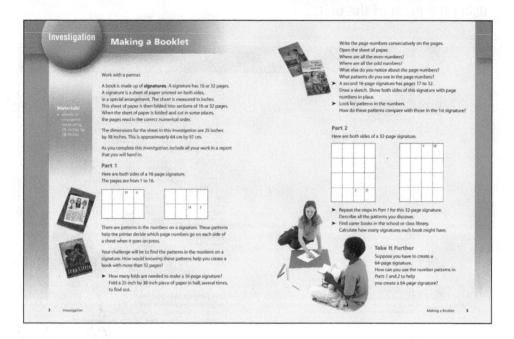

Icons remind you to use **technology**.
Follow the instructions for using a computer
or calculator to do math.

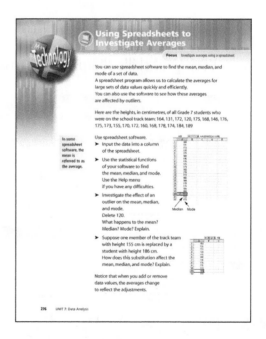

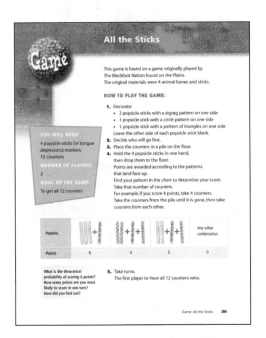

Play a **Game** with your classmates or at home to reinforce your skills.

The World of Work describes how people use mathematics in their careers.

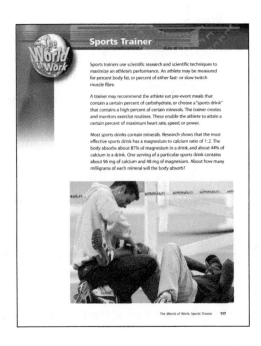

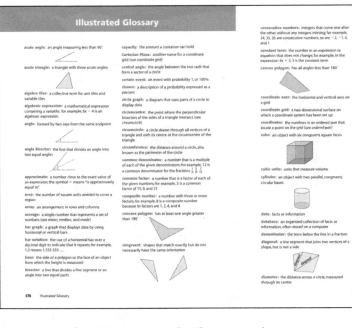

The **Illustrated Glossary** is a dictionary of important math words.

Investigation

Making a Booklet

Work with a partner.

A book is made up of **signatures**. A signature has 16 or 32 pages.
A signature is a sheet of paper printed on both sides,
in a special arrangement. The sheet is measured in inches.
This sheet of paper is then folded into sections of 16 or 32 pages.
When the sheet of paper is folded and cut in some places,
the pages read in the correct numerical order.

Materials:
- sheets of newsprint measuring 25 inches by 38 inches

The dimensions for the sheet in this *Investigation* are 25 inches
by 38 inches. This is approximately 64 cm by 97 cm.

As you complete this *Investigation*, include all your work in a report
that you will hand in.

Part 1

Here are both sides of a 16-page signature.
The pages are from 1 to 16.

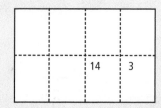

There are patterns in the numbers on a signature. These patterns
help the printer decide which page numbers go on each side of
a sheet when it goes on press.

Your challenge will be to find the patterns in the numbers on a
signature. How would knowing these patterns help you create a
book with more than 32 pages?

➤ How many folds are needed to make a 16-page signature?
Fold a 25-inch by 38-inch piece of paper in half, several times,
to find out.

Write the page numbers consecutively on the pages.
Open the sheet of paper.
Where are all the even numbers?
Where are all the odd numbers?
What else do you notice about the page numbers?
What patterns do you see in the page numbers?

➤ A second 16-page signature has pages 17 to 32.
Draw a sketch. Show both sides of this signature with page numbers in place.

➤ Look for patterns in the numbers.
How do these patterns compare with those in the 1st signature?

Part 2

Here are both sides of a 32-page signature.

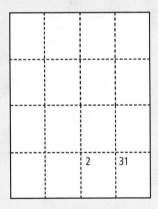

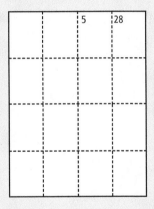

➤ Repeat the steps in *Part 1* for this 32-page signature.
Describe all the patterns you discover.

➤ Find some books in the school or class library.
Calculate how many signatures each book might have.

Take It Further

Suppose you have to create a
64-page signature.
How can you use the number patterns in
Parts 1 and *2* to help
you create a 64-page signature?

Patterns and Relations

Students in a Grade 7 class were raising money for charity.
Some students had a "bowl-a-thon."

This table shows the money that one student raised for different bowling times.

Time (h)	Money Raised ($)
1	8
2	16
3	24
4	32
5	40
6	48

- What patterns do you see in the table?
- Extend the table. For how long would the student have to bowl to raise $72?

What You'll Learn

- Use patterns to explore divisibility rules.
- Translate between patterns and equivalent linear relations.
- Evaluate algebraic expressions by substitution.
- Represent linear relations in tables and graphs.
- Solve simple equations, then verify the solutions.

Why It's Important

- Divisibility rules help us find the factors of a number.
- Graphs provide information and are a useful problem-solving tool.
- Efficient ways to represent a pattern can help us describe and solve problems.

Key Words

- divisibility rules
- algebraic expression
- numerical coefficient
- constant term
- relation
- linear relation
- unit tile
- variable tile
- algebra tiles

1.1 Patterns in Division

Which of these numbers are divisible by 2? By 5? By 10?
How do you know?

- 78
- 27
- 35
- 410

- 123
- 2100
- 4126
- 795

Explore

You will need a hundred chart numbered 301–400, and three different coloured markers.

➤ Use a marker. Circle all numbers on the hundred chart that are divisible by 2.
Use a different marker.
Circle all numbers that are divisible by 4.
Use a different marker.
Circle all numbers that are divisible by 8.
Describe the patterns you see in the numbers you circled.

➤ Choose 3 numbers greater than 400.
Which of your numbers do you think are divisible by 2? By 4? By 8?
Why do you think so?

Reflect & Share

Share your work with another pair of classmates.
Suppose a number is divisible by 8.
What else can you say about the number?
Suppose a number is divisible by 4.
What else can you say about the number?

Connect

We know that 100 is divisible by 4: $100 \div 4 = 25$
So, any multiple of 100 is divisible by 4.
To find out if any whole number with 3 or more digits is divisible by 4, we only need to check the last 2 digits.

To find out if 352 is divisible by 4, check if 52 is divisible by 4.

$52 \div 4 = 13$

52 is divisible by 4, so 352 is divisible by 4.

To check if a number, such as 1192, is divisible by 8,

think: $1192 = 1000 + 192$

We know 1000 is divisible by 8: $1000 \div 8 = 125$

So, we only need to check if 192 is divisible by 8.

Use mental math. $192 \div 8 = 24$

192 is divisible by 8, so 1192 is divisible by 8.

All multiples of 1000 are divisible by 8.

So, for any whole number with 4 or more digits,

we only need to check the last 3 digits to find out if the

number is divisible by 8.

> Another way to check if a number is divisible by 8 is to divide by 4. If the quotient is even, then the number is divisible by 8.

A number that is divisible by 8 is also divisible by 2 and by 4

because $8 = 2 \times 4$.

So, a number divisible by 8 is even.

> 2 and 4 are factors of 8.

You can use patterns to find **divisibility rules**
for other numbers.

➤ All multiples of 10, such as 30, 70, and 260,
  end in 0.

  Any number whose ones digit is 0,
  is divisible by 10.

➤ Here are some multiples of 5.
  5, 10, 15, 20, 25, 30, 35, 40, …, 150, 155, 160, …
  The ones digits form a repeating pattern.
  The core of the pattern is: 5, 0

  Any number whose ones digit is 0 or 5,
  is divisible by 5.

> Every multiple of 5 has a ones digit of 0 or 5.

➤ Multiples of 2 are even numbers: 2, 4, 6, 8, 10, …
  All even numbers are divisible by 2.

  Any number whose ones digit is even, is divisible by 2.

1	2	3	4	5	6	7	8	9	10
11	12	13	14	15	16	17	18	19	20
21	22	23	24	25	26	27	28	29	30
31	32	33	34	35	36	37	38	39	40
41	42	43	44	45	46	47	48	49	50
51	52	53	54	55	56	57	58	59	60
61	62	63	64	65	66	67	68	69	70
71	72	73	74	75	76	77	78	79	80
81	82	83	84	85	86	87	88	89	90
91	92	93	94	95	96	97	98	99	100

Example

Which numbers are divisible by 5? By 8? Both by 5 and by 8?
How do you know?
12, 24, 35, 56, 80, 90, 128, 765, 1048, 1482, 3960, 15 019

A Solution

Any number with 0 or 5 in the ones place is divisible by 5.
So, the numbers divisible by 5 are: 35, 80, 90, 765, 3960

The divisibility rule for 8 only applies when a number is 1000 or greater.
For numbers less than 1000, use mental math or a calculator.
All multiples of 8 are even, so reject 35, 765, and 15 019.
Use mental math to identify that 12 and 90 are not divisible by 8.
Use mental math to identify that 24, 56, 80, and 128 are divisible by 8.
1048 and 3960 are divisible by 8 because 48 and 960 are divisible by 8.
1482 is not divisible by 8 because 482 is not divisible by 8.

We can display the results in a Venn diagram.

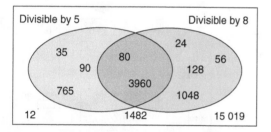

The numbers in the overlapping region are divisible both by 5 and by 8.
So, 80 and 3960 are also divisible by 40, since 5 × 8 = 40.

Practice

1. Which numbers are divisible by 2? By 5?
How do you know?
  a) 106   **b)** 465   **c)** 2198
  d) 215   **e)** 1399   **f)** 4530

2. Explain why a number with 0 in the ones place is divisible by 10.

3. Which numbers are divisible by 4? By 8? By 10?
   How do you know?
   a) 212   **b)** 512   **c)** 5450
   d) 380   **e)** 2132   **f)** 12 256

4. Maxine and Tony discuss divisibility.
   Maxine says, "260 is divisible by 4 and by 5.
   $4 \times 5 = 20$, so 260 is also divisible by 20."
   Tony says, "148 is divisible by 2 and by 4.
   $2 \times 4 = 8$, so 148 is also divisible by 8."
   Are both Maxine and Tony correct? Explain your thinking.

5. Write 3 numbers that are divisible by 8.
   How did you choose the numbers?

6. 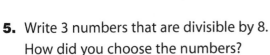 **Assessment Focus**
   a) Use the divisibility rules for 2, 4, and 8 to sort these numbers.

1046	322	460	1784	28
54	1088	224	382	3662

   b) Draw a Venn diagram with 3 loops.
   Label the loops: "Divisible by 2," "Divisible by 4," and "Divisible by 8"
   Explain why you drew the loops the way you did.
   Place the numbers in part a in the Venn diagram.
   How did you decide where to place each number?
   c) Find and insert 3 more 4-digit numbers in the Venn diagram.

7. Use the digits 0 to 9. Replace the ☐ in each number to make
   a number divisible by 4. Find as many answers as you can.
   a) 822☐   **b)** 211 4☐8   **c)** 15 ☐32

8. Take It Further A leap year occurs every 4 years.
   The years 1992 and 2004 were leap years.
   What do you notice about these numbers?
   Was 1964 a leap year? 1852? 1788? Explain.

Reflect

Compare the divisibility rules for 4 and 8.
How can you use one rule to help you remember the other?

Division can be thought of
as making equal groups.

For 20 ÷ 4, we make 4 equal groups of 5.

Explore

 Use a calculator.

➤ Choose 10 different numbers.
Divide each number by 0.
What do you notice?
What do you think this means?

➤ Choose 15 consecutive 2-digit numbers.
Divide each number by 3 and by 9.
Repeat for 15 consecutive 3-digit numbers.

List the numbers that were divisible by 3 and by 9.
Find the sum of the digits of each number.
What do you notice?

Choose 4 different 4-digit numbers you think are divisible by 3 and by 9.
Divide each number by 3 and by 9 to check.
Add the digits in each number. What do you notice?

➤ Draw this Venn diagram.
Sort these numbers.
12　21　42　56　88　135　246　453　728
What can you say about the numbers
in the overlapping region?

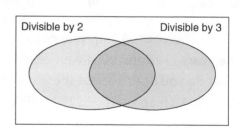

Reflect & Share

Share your work with another pair of classmates.
Explain how to choose a 4-digit number that is divisible by 3.
Without dividing, how can you tell if a number is divisible by 6? By 9?
Why do you think a number cannot be divided by 0?

We can use divisibility rules to find the factors of a number, such as 100.

Any number is divisible by 1 and itself,
so 1 and 100 are factors of 100.

100 is even, so 100 is divisible by 2.

We know 100 is divisible by 4.

The ones digit is 0, so 100 is divisible
by 5 and by 10.

100 is not divisible by 3, 6, 8, or 9.

The factors of 100, from least to greatest, are:

1, 2, 4, 5, 10, 20, 25, 50, 100

$$100 \div 1 = 100$$
$$100 \div 2 = 50$$
$$100 \div 4 = 25$$
$$100 \div 5 = 20$$
$$100 \div 10 = 10$$

Factors occur in pairs.

When we find one factor of a number, we also find a second factor.

A whole number cannot be divided by 0.

We cannot take a given number and share it into zero equal groups.

We cannot make sets of zero from a given number of items.

Example

Edward has 16 souvenir miniature hockey sticks.
He wants to share them equally among his cousins.
How many sticks will each cousin get if Edward has:

a) 8 cousins?   **b)** 0 cousins?

Explain your answer to part b.

A Solution

a) There are 16 sticks. Edward has 8 cousins.

$16 \div 8 = 2$

Each cousin will get 2 sticks.

b) There are 16 sticks. Edward has no cousins.

16 sticks cannot be shared equally among no cousins.

This answer means that we cannot divide a number by zero.

We cannot divide 16 by 0 because 16 cannot be shared
into zero equal groups.

You have sorted numbers in a Venn diagram. You can also use a *Carroll diagram* to sort numbers.

Here is an example:

	Divisible by 3	Not Divisible by 3
Divisible by 8	24, 120, 1104, 12 096	32, 224, 2360
Not Divisible by 8	12, 252, 819, 11 337	10, 139, 9212

Divisibility Rules

A whole number is divisible by:

2 if the number is even

3 if the sum of the digits is divisible by 3

4 if the number represented by the last 2 digits is divisible by 4

5 if the ones digit is 0 or 5

6 if the number is divisible by 2 and by 3

8 if the number represented by the last 3 digits is divisible by 8

9 if the sum of the digits is divisible by 9

10 if the ones digit is 0

Practice

1. Which numbers are divisible by 3? By 9? How do you know?
   a) 117   b) 216   c) 4125   d) 726   e) 8217   f) 12 024

2. Write 3 numbers that are divisible by 6. How did you choose the numbers?

3. Which of these numbers is 229 344 divisible by? How do you know?
   a) 2   b) 3   c) 4   d) 5   e) 6   f) 8   g) 9   h) 10

4. Use the divisibility rules to find the factors of each number.
   How do you know you have found all the factors?
   a) 150   b) 95   c) 117   d) 80

5. Use a Carroll diagram.
   Which numbers are divisible by 4? By 9? By 4 and by 9? By neither 4 or 9?
   144   128   252   153   235   68   120   361   424   468

6. I am a 3-digit number that has a 2 in the hundreds place.
I am divisible by 3, 4, and 5. Which number am I?

7. **Assessment Focus**
  a) Write a 3-digit number that is divisible by 5 and by 9.
     How did you choose the number?
  b) Find the factors of the number in part a. Use the divisibility rules to help you.
  c) How would you find the greatest 3-digit number that is divisible
     by 5 and by 9? The least 3-digit number? Explain your methods.

8. Use the digits 0 to 9.
Replace the □ in each number to make a number divisible by 3.
Find as many answers as you can.
  a) 4□6   b) 1□32   c) 24 71□

9. Suppose you have 24 cereal bars.
You must share the bars equally with everyone in the classroom.
How many cereal bars will each person get, in each case?
  a) There are 12 people in the classroom.
  b) There are 6 people in the classroom.
  c) There is no one in the classroom.
  d) Use your answer to part c.
     Explain why a number cannot be divided by 0.

10. Take It Further Universal Product Codes (UPCs) are used to identify retail products.
The codes have 12 digits, and sometimes start with 0.
To check that a UPC is valid, follow these steps:
  • Add the digits in the odd-numbered positions (1st, 3rd, 5th,…).
  • Multiply this sum by 3.
  • To this product, add the digits in the even-numbered positions.
  • The result should be a number divisible by 10.
Look at this UPC. Is it a valid code? Explain.
Find 2 UPC labels on products at home.
Check to see if the codes are valid. Record your results.

Reflect

Which divisibility rules do you find easiest to use?
Which rules do you find most difficult? Justify your choices.

Writing to Explain Your Thinking

Have you ever tried to explain how you solved a problem to a classmate?

Communicating your thinking can be difficult.

A *Thinking Log* can be used to record what and how you are thinking as you solve a problem.
It is a good way to organize your thoughts.

A classmate, teacher, or parent should be able to follow your thinking to understand how you solved the problem.

Using a Thinking Log

Complete a Thinking Log as you work through this problem.

Nine players enter the Saskatchewan Thumb Wrestling Championship.
In the first round, each player wrestles every other player once.
How many matches are there in the first round?

Thinking Log   Name:_____

I have been asked to find . . .

Here's what I'll try first . . .

To solve this problem I'll . . .

And then . . .

And then . . .

Here's my solution . . .

Reflect & Share

- Read over what you have written.
  Will someone else be able to follow your thinking?
- Share your Thinking Log with a classmate.
  Was your classmate able to follow your thinking and
  understand your solution? Explain.
- Describe any changes you would make to improve your
  Thinking Log.

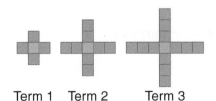

Complete a Thinking Log for each of these problems.

1. I am a 2-digit number. I have three factors.
I am divisible by five. Which number am I?

2. A book contains 124 pages numbered from 1 to 124.
How many times does the digit 7 appear?

3. Here is a pattern of tiles.

Term 1   Term 2   Term 3

a) How many tiles will there be in the 10th term?
b) Which term has 37 tiles? How do you know?

1.3   Algebraic Expressions

We can use symbols to represent a pattern.

Explore

Tehya won some money in a competition.
She has two choices as to how she gets paid.
Choice 1: $20 per week for one year
Choice 2: $400 cash now plus $12 per week for one year

Which method would pay Tehya more money?
For what reasons might Tehya choose each method of payment?

Reflect & Share

Work with another pair of classmates.
For each choice, describe a rule you can use to calculate the total money
Tehya has received at any time during the year.

Connect

We can use a variable to represent a number in an expression.
For example, we know there are 100 cm in 1 m.

1 m = 100 cm

We can write 1×100 cm in 1 m.
There are 2×100 cm in 2 m.
There are 3×100 cm in 3 m.

> Recall that a variable is a letter, such as n, that represents a quantity that can vary.

To write an expression for the number of centimetres in any number of metres,
we say there are $n \times 100$ cm in n metres.
n is a variable.
n represents any number we choose.

We can use any letter, such as n or x, as a variable.
The expression $n \times 100$ is written as $100n$.
$100n$ is an **algebraic expression**.

> Variables are written in italics so they are not confused with units of measurement.

Here are some other algebraic expressions, and their meanings.
In each case, n represents the number.

- Three more than a number: $3 + n$ or $n + 3$
- Seven times a number: $7n$
- Eight less than a number: $n - 8$
- A number divided by 20: $\frac{n}{20}$

$7n$ means $7 \times n$.

When we replace a variable with a number in an algebraic
expression, we *evaluate* the expression. That is, we find the value of
the expression for a particular value of the variable.

Example

Write each algebraic expression in words.
Then evaluate for the value of the variable given.

a) $5k + 2$ for $k = 3$ **b)** $32 - \frac{x}{4}$ for $x = 20$

A Solution

a) $5k + 2$ means 5 times a number, then add 2.
Replace k with 3 in the expression $5k + 2$.
Then use the order of operations.

$$5k + 2 = 5 \times 3 + 2 \quad \text{Multiply first.}$$
$$= 15 + 2 \quad \text{Add.}$$
$$= 17$$

b) $32 - \frac{x}{4}$ means 32 minus a number divided by 4.
Replace x with 20 in the expression $32 - \frac{x}{4}$.
Then use the order of operations.

$\frac{x}{4}$ means $x \div 4$.

$$32 - \frac{x}{4} = 32 - \frac{20}{4} \quad \text{Divide first.}$$
$$= 32 - 5 \quad \text{Subtract.}$$
$$= 27$$

In the expression $5k + 2$,

- 5 is the **numerical coefficient** of the variable.
- 2 is the **constant term**.
- k is the *variable*.
  The variable represents any number in a set of numbers.

Practice

1. Identify the numerical coefficient, the variable, and the constant term in each algebraic expression.
   a) $3x + 2$ b) $5n$ c) $w + 3$ d) $2p + 4$

2. An algebraic expression has variable p, numerical coefficient 7, and constant term 9.
   Write as many different algebraic expressions as you can that fit this description.

3. Write an algebraic expression for each phrase.
   a) six more than a number
   b) a number multiplied by eight
   c) a number decreased by six
   d) a number divided by four

4. A person earns $4 for each hour he spends baby-sitting.
   a) Find the money earned for each time.
      i) 5 h   ii) 8 h
   b) Write an algebraic expression you could use to find the money earned in t hours.

5. Write an algebraic expression for each sentence.
   a) Double a number and add three.
   b) Subtract five from a number, then multiply by two.
   c) Divide a number by seven, then add six.
   d) A number is subtracted from twenty-eight.
   e) Twenty-eight is subtracted from a number.

6. a) Write an algebraic expression for each phrase.
      i) four more than a number
      ii) a number added to four
      iii) four less than a number
      iv) a number subtracted from four
   b) How are the expressions in part a alike?
      How are they different?

7. Evaluate each expression by replacing *x* with 4.
   a) $x + 5$
   b) $3x$
   c) $2x - 1$
   d) $\frac{x}{2}$
   e) $3x + 1$
   f) $20 - 2x$

8. Evaluate each expression by replacing *z* with 7.
   a) $z + 12$
   b) $10 - z$
   c) $5z$
   d) $3z - 3$
   e) $35 - 2z$
   f) $3 + \frac{z}{7}$

9. **Assessment Focus** Jason works at a local fish and chips restaurant.
   He earns \$7/h during the week, and \$9/h on the weekend.

   a) Jason works 8 h during the week and 12 h on the weekend.
      Write an expression for his earnings.
   b) Jason works *x* hours during the week and 5 h on the weekend.
      Write an expression for his earnings.
   c) Jason needs \$115 to buy sports equipment.
      He worked 5 h on the weekend.
      How many hours does Jason have to work during the week to have the money he needs?

10. **Take It Further** A value of *n* is substituted in each expression to get the number in the box.
   Find each value of *n*.
   a) $5n$ | 30
   b) $3n - 1$ | 11
   c) $4n + 7$ | 15
   d) $5n - 4$ | 11
   e) $4 + 6n$ | 40
   f) $\frac{n}{8}$ | 5

Reflect

Explain why it is important to use the order of operations when evaluating an algebraic expression.
Use an example in your explanation.

Here is a pattern made from linking cubes.

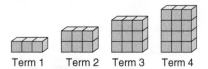

Term 1   Term 2   Term 3   Term 4

A pattern rule is: Start at 3. Add 3 each time.
This rule relates each term to the term that comes before it.

We can also describe this pattern using the term number.

Term Number	1	2	3	4
Term	3	6	9	12

How does each term relate to the term number?

Explore

On Enviro-Challenge Day, Grade 7 classes compete
to see which class can collect the most garbage.

Each student in Ms. Thomson's class pledges
to pick up 6 pieces of garbage.
➤ How many pieces of garbage will be picked up when
   the number of students is 5? 10? 15? 20? 25? 30?
➤ What pattern do you see in the numbers of pieces
   of garbage?
➤ Write a rule to find how many pieces of garbage
   will be picked up, when you know the number of students.
➤ Write an algebraic expression for the number of pieces
   of garbage picked up by *n* students.

Reflect & Share

Share your work with another pair of classmates.
Find the number of pieces of garbage picked up by 35 students.
How can you do this using the pattern?
Using the rule? Using the algebraic expression?

Miss Jackson's class pledges to pick up a total of 10 more pieces
of garbage than Ms. Thomson's class.
Here are the numbers of pieces of garbage picked up by
different numbers of students.

Number of students	2	4	6	8	10	12
Number of pieces of garbage picked up by Ms. Thomson's class	12	24	36	48	60	72
Number of pieces of garbage picked up by Miss Jackson's class	22	34	46	58	70	82

Pieces of garbage
picked up by   =   10   +   Pieces of garbage
Miss Jackson's class   picked up by
                                          Ms. Thomson's class

Let n represent the number of students who pick up garbage in Ms. Thomson's class.
Then the number of pieces of garbage picked up by Ms. Thomson's class is $6n$.
And, the number of pieces of garbage picked up
by Miss Jackson's class is $10 + 6n$.

> Recall that 10 is
> the constant term.

The number of pieces of garbage is *related* to the number of students.
When we compare or *relate* a variable to an expression that contains
the variable, we have a **relation**.
That is, $10 + 6n$ is related to n.   **This is a relation.**

Example

Mr. Prasad plans to hold a party for a group of his friends.
The cost of renting a room is $35.
The cost of food is $4 per person.

a) Write a relation for the cost of the party, in dollars,
   for n people.
b) How much will a party cost for 10 people?
   For 15 people?
c) How does the relation change if the cost of food doubles?
   How much more would a party for 10 people cost?
   How do you know the answer makes sense?

A Solution

a) The cost of renting a room is $35.

This does not depend on how many people come.

The cost of food is $4 per person.

If 5 people come, the cost of food in dollars is: $4 \times 5 = 20$

If n people come, the cost of food in dollars is: $4 \times n$, or $4n$

So, n is related to $35 + 4n$.

b) To find the cost for 10 people, substitute $n = 10$ into $35 + 4n$.

$$35 + 4n = 35 + 4(10)$$
$$= 35 + 40$$
$$= 75$$

The party will cost $75.

To find the cost for 15 people, substitute $n = 15$ into $35 + 4n$.

$$35 + 4n = 35 + 4(15)$$
$$= 35 + 60$$
$$= 95$$

The party will cost $95.

> $4(10)$ means 4×10.

c) If the cost of food doubles, Mr. Prasad will pay $8 per person.

If n people come, the cost for food, in dollars, is $8n$.

For n people, the cost of the party, in dollars, is now $35 + 8n$.

If 10 people come, the cost is now:

$$35 + 8n = 35 + 8(10)$$
$$= 35 + 80$$
$$= 115$$

The party will cost $115.

This is an increase of $115 - $75 = $40.

The answer makes sense because the cost is now $4 more per person.

So, the extra cost for 10 people would be 4×10, or $40 more.

Math Link

History

The word "algebra" comes from the Arabic word "al-jabr." This word appeared in the title of one of the earliest algebra texts, written around the year 825 by al-Khwarizmi. He lived in what is now Uzbekistan.

Practice

1. i) For each number pattern, how is each term related to the term number?
   ii) Let n represent any term number. Write a relation for the term.

a)

Term Number	1	2	3	4	5	6
Term	2	4	6	8	10	12

b)

Term Number	1	2	3	4	5	6
Term	3	4	5	6	7	8

c)

Term Number	1	2	3	4	5	6
Term	8	16	24	32	40	48

d)

Term Number	1	2	3	4	5	6
Term	6	7	8	9	10	11

2. There are n students in a class. Write a relation for each statement.
   a) the total number of pencils, if each student has three pencils
   b) the total number of desks, if there are two more desks than students
   c) the total number of geoboards, if each pair of students shares one geoboard
   d) the total number of stickers, if each student gets four stickers and there are ten stickers left over

3. A person earns \$10 for each hour worked.
   a) Write a relation for her earnings for n hours of work.
   b) How much does she earn for 30 h of work?

4. a) Write a relation for the perimeter of a square with side length n centimetres.
   b) What is the perimeter of a square with side length 12 cm?
   c) Suggest a situation that could be represented by each relation.
      i) $3s$ is related to s ii) $8t$ is related to t

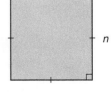

5. Suggest a real-life situation that could be represented by each relation.
   a) $n + 5$ is related to n b) $15 + 2p$ is related to p
   c) $3t + 1$ is related to t
   How do you know each situation fits the relation?

6. Koko is organizing an overnight camping trip. The cost to rent a campsite is $20. The cost of food is $9 per person.

   a) How much will the trip cost if 5 people go? 10 people go?
   b) Write a relation for the cost of the trip when *p* people go.
   c) Suppose the cost of food doubles.
      Write a relation for the total cost of the trip for *p* people.
   d) Suppose the cost of the campsite doubles.
      Write a relation for the total cost of the trip for *p* people.
   e) Explain why using the variable *p* is helpful.

7. **Assessment Focus**   A pizza with cheese and tomato toppings costs $8.00. It costs $1 for each extra topping.

   a) Write a relation for the cost of a pizza with *e* extra toppings.
   b) What is the cost of a pizza with 5 extra toppings?
   c) On Tuesdays, the cost of the same pizza with cheese and tomato toppings is $5.00. Write a relation for the cost of a pizza with *e* extra toppings on Tuesdays.
   d) What is the cost of a pizza with 5 extra toppings on Tuesdays?
   e) How much is saved by buying the pizza on Tuesday?

8. Write a relation for the pattern rule for each number pattern. Let *n* represent any term number.
   a) 4, 8, 12, 16, …   **b)** 7, 8, 9, 10, …   **c)** 0, 1, 2, 3, …

9. **Take It Further**
   i) For each number pattern, how is each term related to the term number?
   ii) Let *n* represent any term number. Write a relation for the term.

a)

Term Number	1	2	3	4	5	6
Term	3	5	7	9	11	13

b)

Term Number	3	4	5	6	7	8
Term	7	10	13	16	19	22

c)

Term Number	2	3	4	5	6	7
Term	5	9	13	17	21	25

Reflect

How did your knowledge of patterning help you in this lesson?

An Input/Output machine represents a relation.
Any Input number can be represented by *n*.

Suppose you input *n* = 8.
What will the output be?
How is the output related to the input?

Explore

Sketch an Input/Output machine like this one.

Write an algebraic expression to go in the machine.

➤ Use the numbers 1 to 6 as input.
  Find the output for each Input number.
  Record the input and output in a table
  like this.

➤ How is the output related to the input?

➤ Describe the pattern in the Output numbers.

Input	Output
1	
2	
3	
4	
5	
6	

Reflect & Share

Share your work with another pair of classmates.
Describe how you would find the next 3 Output numbers
for your classmates' Input/Output machine.
How is the output related to the input?

This Input/Output machine relates n and $2n + 3$.
To create a table of values,
select a set of Input numbers.
To get each Output number, multiply the
Input number by 2, then add 3.

When $n = 1, 2n + 3 = 2(1) + 3$
$= 2 + 3$
$= 5$

When $n = 2, 2n + 3 = 2(2) + 3$
$= 4 + 3$
$= 7$

When $n = 3, 2n + 3 = 2(3) + 3$
$= 6 + 3$
$= 9,$

Remember the *order of operations*. Multiply before adding.

Input n	Output $2n + 3$
1	5
2	7
3	9
4	11
5	13

and so on.

We used consecutive Input numbers.
The Output numbers form a pattern. They increase by 2 each time.
This is because the expression contains $2n$,
which means that the Input number is doubled.
When the Input number increases by 1, the Output number increases by 2.

The expression $2n + 3$ can also be written as $3 + 2n$.

When a relation is represented as a table of values,
we can write the relation using algebra.

Example

Write the relation represented by this table.

Input	Output
1	2
2	5
3	8
4	11
5	14

A Solution

Let any Input number be represented by n.
The input increases by 1 each time.
The output increases by 3 each time.
This means that the expression for the output contains $3n$.

Input	Output
1	2
2	5
3	8
4	11
5	14

$+1$ (between inputs) $\qquad +3$ (between outputs)

Substitute several values of n in $3n$, then look for a pattern.

When $n = 1, 3n = 3(1) = 3$
When $n = 2, 3n = 3(2) = 6$
When $n = 3, 3n = 3(3) = 9$
When $n = 4, 3n = 3(4) = 12$
When $n = 5, 3n = 3(5) = 15$

Each value is 1 more than the output above. That is, the output is 1 less than each value.

So, the output is $3n - 1$.
The table shows how $3n - 1$ relates to n.

Another Solution

Another way to solve this problem is to notice that each output is 1 less than a multiple of 3.
So, the output is $3 \times n - 1$, or $3n - 1$.
The table shows how $3n - 1$ relates to n.

Input	Output
1	$2 = 3 \times 1 - 1$
2	$5 = 3 \times 2 - 1$
3	$8 = 3 \times 3 - 1$
4	$11 = 3 \times 4 - 1$
5	$14 = 3 \times 5 - 1$
n	$3 \times n - 1$

Practice

1. Copy and complete each table.
Explain how the Output number is related to the Input number.

a)

Input x	Output $2x$
1	
2	
3	
4	
5	

b)

Input m	Output $10 - m$
1	
2	
3	
4	
5	

c)

Input p	Output $3p + 5$
1	
2	
3	
4	
5	

2. Use algebra. Write a relation for each Input/Output table.

a)

Input n	Output
1	7
2	14
3	21
4	28

b)

Input n	Output
1	4
2	7
3	10
4	13

c)

Input n	Output
1	1
2	3
3	5
4	7

3. **Assessment Focus** For each table, find the output.
Explain how the numbers 3 and 4 in each relation affect the output.

a)

Input n	Output $3n + 4$
1	
2	
3	
4	

b)

Input n	Output $4n + 3$
1	
2	
3	
4	

4. Use algebra. Write a relation for each Input/Output table.

a)

Input x	Output
1	5
2	8
3	11
4	14

b)

Input x	Output
1	1
2	7
3	13
4	19

c)

Input x	Output
1	8
2	13
3	18
4	23

5. Take It Further
   a) Describe the patterns in this table.
   b) Use the patterns to extend
      the table 3 more rows.
   c) Use algebra.
      Write a relation that describes
      how the output is related to the input.

Input x	Output
5	1
15	3
25	5
35	7
45	9
55	11

Reflect

Your friend missed today's lesson. Explain how to write the
relation represented by an Input/Output table.

Mid-Unit Review

1.1 **1.** Which numbers are divisible by 4? By 8? How do you know?
   a) 932   b) 1418   c) 5056
   d) 12 160   e) 14 436

1.2 **2.** Draw a Venn diagram with 2 loops. Label the loops: "Divisible by 3" and "Divisible by 5." Sort these numbers: 54 85 123 735 1740 3756 6195 What is true about the numbers in the overlapping region?

3. Use the divisibility rules. Find the factors of each number.
   a) 85   b) 136   c) 270

1.3 **4.** Write an algebraic expression for each statement. Let n represent the number.
   a) seven more than a number
   b) a number multiplied by eleven
   c) a number divided by six
   d) three less than four times a number
   e) the sum of two and five times a number

1.4 **5.** Predict which expression in each pair will have the greater value when y is replaced with 8. Evaluate to check your predictions.
   a) i) $y + 7$ ii) $2y$
   b) i) $6y$ ii) $9 - y$
   c) i) $\frac{y + 4}{2}$ ii) $\frac{y}{2} + 4$
   d) i) $2y + 6$ ii) $3y - 6$

6. i) For each number pattern, how is each term related to the term number?
   ii) Let n represent the term number. Write a relation for the term.

   a)

Term Number	1	2	3	4	5	6
Term	6	12	18	24	30	36

   b)

Term Number	1	2	3	4	5	6
Term	5	6	7	8	9	10

7. Dave pays to practise in a music studio. He pays $12 each month, plus $2 for each hour he practises.
   a) Write a relation for the total cost for one month, in dollars, when Dave practises t hours.
   b) How much will Dave pay to practise 10 h in one month? 20 h?
   c) How does the relation change when the cost per hour doubles?

1.5 **8.** Use algebra. Write a relation for each Input/Output table.

   a)

Input x	Output
1	7
2	11
3	15
4	19

   b)

Input x	Output
1	5
2	13
3	21
4	29

We can use a graph to show the relationship between two quantities.
What does this graph show?

How many jellybeans are in each bag?
Write a relation for the total number of jellybeans in *n* bags.

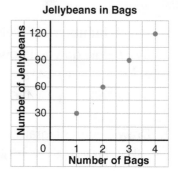

Jellybeans in Bags

Explore

You will need grid paper.

The cost of *n* CDs, in dollars, is 12*n*.
➤ What is the cost of one CD?
➤ Copy and complete this table.
➤ Graph the data.

Use the graph to answer these questions:
➤ What is the cost of 5 CDs?
➤ How many CDs could you buy with $72?

Number of CDs n	Cost ($) $12n$
0	
2	
4	
6	
8	
10	

Reflect & Share

Describe the patterns in the table. How are these patterns shown in the graph?
If you had $50, how many CDs could you buy?

This table shows how $4n + 2$ relates to n, where n is a whole number.

We could have chosen any Input numbers, but to see patterns it helps to use consecutive numbers.

These data are plotted on a graph.
The input is plotted on the horizontal axis.
The output is plotted on the vertical axis.
On the vertical axis, the scale is 1 square for every 2 units.
The graph also shows how $4n + 2$ relates to n.

Input n	Output $4n + 2$
0	$4(0) + 2 = 2$
1	$4(1) + 2 = 6$
2	$4(2) + 2 = 10$
3	$4(3) + 2 = 14$
4	$4(4) + 2 = 18$
5	$4(5) + 2 = 22$

+ 1 (between each Input row)
+ 4 (between each Output row)

When we place a ruler along the points, we see the graph is a set of points that lie on a straight line.
When points lie on a straight line, we say the relation is a **linear relation**.
Since no numbers lie between the Input values in the table, it is not meaningful to join the points with a solid line.

The graph shows that each time the input increases by 1, the output increases by 4.

Graph of $4n + 2$ against n

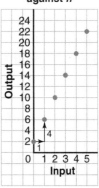

Example

Mr. Beach has 25 granola bars.
He gives 3 granola bars to each student who stays after school to help prepare for the school concert.

a) Write a relation to show how the number of granola bars that remain is related to the number of helpers.

b) Make a table to show this relation.

c) Graph the data. Describe the graph.

d) Use the graph to answer these questions:

   i) How many granola bars remain when 7 students help?

   ii) When will Mr. Beach not have enough granola bars?

A Solution

a) Let n represent the number of helpers.
Each helper is given 3 granola bars.
So, the number of granola bars given to n helpers is $3n$.
There are 25 granola bars.
The number of granola bars that remain is $25 - 3n$.
So, n is related to $25 - 3n$.

b) Substitute each value of n into $25 - 3n$.

Number of Helpers n	Number of Granola Bars Left $25 - 3n$
0	$25 - 3(0) = 25$
1	$25 - 3(1) = 22$
2	$25 - 3(2) = 19$
3	$25 - 3(3) = 16$
4	$25 - 3(4) = 13$
5	$25 - 3(5) = 10$

c) On the vertical axis, use a scale of 1 square for every 2 units.
The points lie on a line so the graph represents a linear relation.
When the input increases by 1, the output decreases by 3.

The graph goes down to the right.
This is because the number of granola bars that remain decreases as the number of helpers increases.

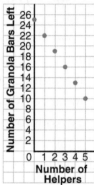

Granola Bars Left

d) i) To find the number of granola bars that remain, extend the graph.
The points lie on a straight line.
Extend the graph to 7 helpers.
There are 4 granola bars left.

ii) Continue to extend the graph.
25 granola bars are enough for 8 helpers, but not for 9 helpers.
Mr. Beach will not have enough granola bars for 9 or more helpers.

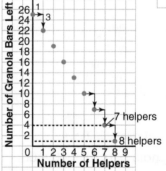

Granola Bars Left

To graph a relation, follow these steps:
- Select appropriate Input numbers. Make a table of values.
- Choose scales for the horizontal and vertical axes.
- Use a ruler to draw the axes on grid paper.
  Use numbers to indicate the scale.
- Label the axes. Give the graph a title.
- Plot the data in the table.

Another Strategy
We could have solved part d of the *Example* by extending the table.

Practice

1. Copy and complete this Input/Output table for each relation.
   a) 4n is related to n.
   b) x + 3 is related to x.
   c) 4c + 6 is related to c.

Input n	Output
1	
2	
3	
4	
5	

2. Graph each relation in question 1.
Suggest a real-life situation it could represent.

3. **a)** Copy and complete this Input/Output table to show how 6a − 4 is related to a.
   b) Graph the relation.
   What scale did you use on the vertical axis?
   How did you make your choice.
   c) Explain how the graph illustrates the relation.

Input a	Output
2	
4	
6	
8	
10	

4. Look at the graph on the right.
   a) What is the output when the input is 1?
   b) Which input gives the output 18?
   c) Extend the graph. What is the output when the input is 8?
   d) Suggest a real-life situation this graph could represent.

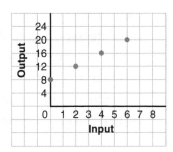

5. Admission to Fun Place is $5.
Each go-cart ride costs an additional $3.
   a) Write a relation to show how the total cost is related to the number of go-cart rides.
   b) Copy and complete this table.
   c) Draw a graph to show the relation.
   Describe the graph.
   d) Use the graph to answer these questions:
      i) Erik goes on 6 go-cart rides.
      What is his total cost?
      ii) Before entering the park, Lydia has $30.
      How many go-cart rides can she afford?

Number of Go-Cart Rides	Total Cost ($)
0	
1	
2	
3	
4	
5	

6. Match each graph to its relation.

   a) The number of seashells collected is related
      to the number of students who collected.
      There are 12 seashells to start.
      Each student collects 3 seashells.

   b) The number of counters on the teacher's desk is related
      to the number of students who remove counters.
      There are 36 counters to start.
      Each student removes 6 counters.

   c) The money earned baby-sitting is related to the number
      of hours worked. The baby-sitter earns $6/h.

i) 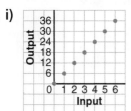   **ii)** 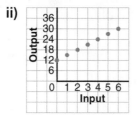   **iii)**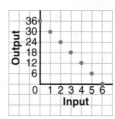

7. Akuti borrows $75 from her mother to buy a new lacrosse stick.
   She promises to pay her mother $5 each week until her debt is paid off.

   a) Write a relation to show how the amount Akuti owes is related to
      the number of weeks.
   b) Make a table for the amount owing after 2, 4, 6, 8, and 10 weeks.
   c) Draw a graph to show the relation. Describe the graph.
   d) Use the graph to answer these questions:
      i) How much does Akuti owe her mother after 13 weeks?
      ii) When will Akuti finish paying off her debt?

8. **Assessment Focus** Use the relation: $5n + 6$ is related to n
   a) Describe a real-life situation that could be represented by this relation.
   b) Make a table of values using appropriate Input numbers.
   c) Graph the relation. Describe the graph.
   d) Write 2 questions you could answer using the graph.
      Answer the questions.

Reflect

How can the graph of a relation help you answer questions
about the relation? Use an example to show your thinking.

Explore

Part 1

➤ Write an algebraic expression for these statements:
Think of a number.
Multiply it by 3.
Add 4.

➤ The answer is 13. What is the original number?

Part 2

➤ Each of you writes your own number riddle.
Trade riddles with your partner.

➤ Write an algebraic expression for your partner's statements.
Find your partner's original number.

Reflect & Share

Compare your answer to *Part 1* with that of another pair of classmates.
If you found different values for the original number, who is correct?
Can both of you be correct? How can you check?

Connect

Zena bought 3 CDs.
Each CD costs the same amount.
The total cost was $36.
What was the cost of 1 CD?

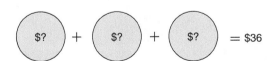

We can write an equation for this situation.
Let p dollars represent the cost of 1 CD.
Then, the cost of 3 CDs is $3p$. This is equal to $36.
We can write an equation to represent this situation:
$3p = 36$

When we write one quantity equal to another quantity, we have an *equation*.
Each quantity may be a number or an algebraic expression.
For example, $3x + 2$ is an algebraic expression; 11 is a number.
When we write $3x + 2 = 11$, we have an equation.
An equation is a statement that two quantities are equal.
Each side of the equation has the same value.

In an equation, the variable represents a specific unknown number.
When we find the value of the unknown number, we *solve* the equation.

Example

Write an equation for each sentence.

a) Three more than a number is 15.
b) Five less than a number is 7.
c) A number subtracted from 5 is 1.
d) A number divided by 3 is 10.
e) Eight added to 3 times a number is 26.

A Solution

a) Three more than a number is 15.
   Let x represent the number.
   Three more than x: $x + 3$
   The equation is: $x + 3 = 15$

b) Five less than a number is 7.
   Let z represent the number.
   Five less than z: $z - 5$
   The equation is: $z - 5 = 7$

c) A number subtracted from 5 is 1.
   Let g represent the number.
   g subtracted from 5: $5 - g$
   The equation is: $5 - g = 1$

d) A number divided by 3 is 10.
   Let j represent the number.
   j divided by 3: $\frac{j}{3}$
   The equation is: $\frac{j}{3} = 10$

e) Eight added to 3 times a number is 26.
   Let h represent the number.
   3 times h: $3h$
   8 added to $3h$: $3h + 8$
   The equation is: $3h + 8 = 26$

Practice

1. Write an equation for each sentence.
   a) Eight more than a number is 12.
   b) Eight less than a number is 12.

2. Write a sentence for each equation.
   a) $12 + n = 19$
   b) $3n = 18$
   c) $12 - n = 5$
   d) $\frac{n}{2} = 6$

3. Write an equation for each sentence.

    a) Six times the number of people in the room is 258.

    b) One-half the number of students in the band is 21.

    c) The area of a rectangle with base 6 cm and height h centimetres is 36 cm^2.

4. The perimeter of a square is 156 cm.
Write an equation you could use to find
the side length of the square.

Recall that perimeter
is the distance around
a shape.

5. The side length of a regular hexagon is 9 cm.
Write an equation you could use to find the
perimeter of the hexagon.

6. Match each equation with the correct sentence.

    a) $n + 4 = 8$ **A.** Four less than a number is 8.

    b) $4n = 8$ **B.** Four more than four times a number is 8.

    c) $n - 4 = 8$ **C.** The sum of four and a number is 8.

    d) $4 + 4n = 8$ **D.** The product of four and a number is 8.

7. Alonso thinks of a number.
He divides the number by 4, then adds 10.
The answer is 14.
Write an equation for the problem.

8. **Assessment Focus**

    a) Write an equation for each sentence.

      i) Five times the number of students is 295.

      ii) The area of a rectangle with base 7 cm and
height h centimetres is 28 cm^2.

     iii) The cost of 2 tickets at x dollars each and
5 tickets at $4 each is $44.

     iv) Bhavin's age 7 years from now will be 20 years old.

    b) Which equation was the most difficult to write? Why?

    c) Write your own sentence, then write it as an equation.

Reflect

Give an example of an algebraic expression and of an equation.
How are they similar? How are they different?

Focus   Use algebra tiles and symbols to solve simple equations.

We can use tiles to represent an expression.
One yellow tile ▢ can represent +1.

We call it a **unit tile**.

We also use tiles to represent variables.
This tile represents *x*. ▭

We call it an *x*-tile, or a **variable tile**.

> A unit tile and a variable tile are collectively **algebra tiles**.

What algebraic expression do these tiles represent?

In this lesson, you will learn how to use tiles to solve equations.
In Unit 6, you will learn other ways to solve equations.

Explore

Alison had $13.
She bought 5 gift bags.
Each bag costs the same amount.
Alison then had $3 left.
How much was each gift bag?

➤ Let *d* dollars represent the cost of 1 gift bag.
  Write an equation to represent the problem.

➤ Use tiles. Solve the equation to find the value of *d*.
  How much was each gift bag?

Reflect & Share

Compare your equation with that of another pair of classmates.
If the equations are different, try to find out why.
Discuss your strategies for using tiles to solve the equation.

Owen collects model cars.
His friend gives him 2 cars.
Owen then has 7 cars.
How many cars did he have at the start?

We can write an equation that we can solve
to find out. Let x represent the number
of cars Owen had at the start.
2 more than x is: $x + 2$
The equation is: $x + 2 = 7$

We can use tiles to solve this equation.
We draw a vertical line in the centre of the page.
It represents the equals sign in the equation.

We arrange tiles on each side of the line to represent
the expression or number on each side of the equation.

We want to get the x-tile on its own.
This is called *isolating the variable*.
When we solve an equation, we must *preserve* the equality.
That is, whatever we do to one side of the equation, we must also do to the other side.

To solve the equation $x + 2 = 7$:

On the left side, put tiles
to represent $x + 2$.

On the right side, put tiles
to represent 7.

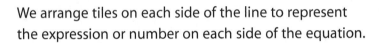

To isolate the x-tile, remove the 2 unit tiles from the left side.
To preserve the equality, remove 2 unit tiles from the right side, too.

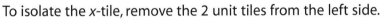

The tiles show the solution is $x = 5$.

To *verify* the solution, replace x with 5 yellow tiles.

Left side: ⬜⬜⬜⬜⬜ + ⬜⬜ ⟶ 7 yellow tiles

Right side: ⬜⬜⬜⬜⬜⬜⬜ ⟶ 7 yellow tiles

Since the left side and right side have equal numbers of tiles,
the solution $x = 5$ is correct.
Owen had 5 cars at the start.

Example

Two more than three times a number is 14.
a) Write an equation you can solve to find the number.
b) Use tiles to solve the equation.
c) Verify the solution.

A Solution

a) Two more than three times a number is 14.
   Let x represent the number.
   Three times x: $3x$
   Two more than $3x$: $3x + 2$
   The equation is: $3x + 2 = 14$
b) $3x + 2 = 14$

Remove 2 unit tiles from each side to isolate the x-tiles.

There are 3 *x*-tiles.
Arrange the tiles remaining on each side into 3 equal groups.

One *x*-tile equals 4 unit tiles.
So, $x = 4$

c) To verify the solution, replace *x* with 4 yellow tiles.

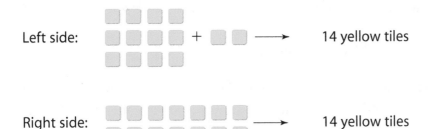

Left side: ⟶ 14 yellow tiles

Right side: ⟶ 14 yellow tiles

Since the left side and right side have equal numbers of tiles,
the solution $x = 4$ is correct.

Practice

Use tiles to solve each equation.
1. Draw pictures to represent the steps you took to solve each equation.
   a) $x + 6 = 13$ b) $4 + x = 12$ c) $11 = x + 7$
   d) $2x = 16$ e) $18 = 3x$ f) $4x = 12$

2. Seven more than a number is 12.
   a) Write an equation for this sentence.
   b) Solve the equation. Verify the solution.

3. For each equation in question 1, identify a constant term,
   the numerical coefficient, and the variable.

4. At the used bookstore, one paperback book costs $3.
How many books can be bought for $12?
  a) Write an equation you can solve to find how many books can be bought.
  b) Solve the equation. Verify the solution.

5. Kiera shared 20 hockey cards equally among her friends.
Each friend had 4 cards.
  a) Write an equation that describes this situation.
  b) Solve the equation to find how many friends shared the cards.

6. In Nirmala's Grade 7 class, 13 students walk to school. There are 20 students in the class.
  a) Write an equation you can solve to find how many students do not walk to school.
  b) Solve the equation. Verify the solution.

7. Jacob is thinking of a number. He multiplies it by 3 and then adds 4. The result is 16.
  a) Write an equation to represent this situation.
  b) Solve the equation to find Jacob's number.

8. **Assessment Focus** Tarana had 2 paper plates. She bought 4 packages of paper plates.
Each package had the same number of plates. Tarana now has a total of 18 plates.
How many paper plates were in each package?
  a) Write an equation you can solve to find how many plates were in each package.
  b) Solve the equation. Verify the solution.

9. **Take It Further** Dominique has 20 comic books. She gives 5 to her sister,
then gives 3 to each of her friends. Dominique has no comic books left.
  a) Write an equation you can solve to find how many friends were given comic books.
  b) Solve the equation. Verify the solution.

10. **Take It Further**
  a) Write an equation whose solution is $x = 4$.
  b) Write a sentence for your equation.
  c) Solve the equation.
  d) Describe a situation that can be represented by your equation.

Reflect

When you solve an equation, how can you be sure that
your solution is correct?

Unit Review

☑ A whole number is divisible by:
- **2** if the number is even
- **3** if the sum of the digits is divisible by 3
- **4** if the number represented by the last 2 digits is divisible by 4
- **5** if the ones digit is 0 or 5
- **6** if the number is divisible by 2 and by 3
- **8** if the number represented by the last 3 digits is divisible by 8
- **9** if the sum of the digits is divisible by 9
- **10** if the ones digit is 0

A whole number cannot be divided by 0.

☑ A *variable* is a letter or symbol.
It represents a set of numbers in an *algebraic expression*.
A variable can be used to write an algebraic expression:
"3 less than a number" can be written as $n - 3$.

A variable represents a number in an *equation*.
A variable can be used to write an equation.
"4 more than a number is 11" can be written as $x + 4 = 11$.

☑ An algebraic expression can be *evaluated* by substituting a number for the variable.
$6r + 3$ when $r = 2$ is: $6 \times 2 + 3 = 12 + 3$
$$= 15$$

☑ A *relation* describes how the output is related to the input.
A relation can be displayed using a table of values or a graph.
When points of a relation lie on a straight line, it is a *linear relation*.

☑ An equation can be solved using tiles.

What Should I Be Able to Do?

LESSON

1.1
1.2
1. Use the divisibility rules to find the factors of 90.

2. Which of these numbers is 23 640 divisible by? How do you know?
a) 2 b) 3 c) 4
d) 5 e) 6 f) 8
g) 9 h) 10 i) 0

3. I am a 3-digit number.
I am divisible by 4 and by 9.
My ones digit is 2.
I am less than 500.
Which number am I?
Find as many numbers as you can.

4. Draw a Venn diagram with 2 loops.
Label the loops "Divisible by 6," and "Divisible by 9."
a) Should the loops overlap? Explain.
b) Write these numbers in the Venn diagram.
330   639   5598   10 217
2295   858   187   12 006
How did you know where to put each number?

1.3
5. i) Write an algebraic expression for each statement.
ii) Evaluate each expression by replacing the variable with 8.
a) five less than a number
b) a number increased by ten
c) triple a number
d) six more than three times a number

1.4
6. There are n women on a hockey team.
Write a relation for each statement.
a) the total number of hockey sticks, if each player has 4 sticks
b) the total number of lockers in the dressing room, if there are 3 more lockers than players
c) the total number of water jugs on the bench, if each group of 4 players shares 1 jug

1.5
7. Copy and complete each table.
Explain how the Output number is related to the Input number.

a)
Input n	Output $n + 13$
1	
2	
3	
4	
5	

b)
Input n	Output $5n + 1$
1	
2	
3	
4	
5	

c)
Input n	Output $6n - 3$
1	
2	
3	
4	
5	

8. Use algebra. Write a relation for each Input/Output table.

a)

Input n	Output
1	12
2	13
3	14
4	15

b)

Input n	Output
1	2
2	7
3	12
4	17

1.6

9. Match each graph with one of the relations below.

a)
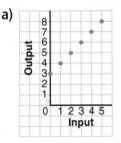

b)
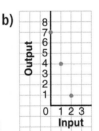

c)
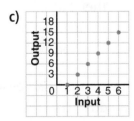

i) $7 - 3n$ is related to n.

ii) $4n + 3$ is related to n.

iii) $n - 1$ is related to n.

iv) $n + 3$ is related to n.

v) $3n - 3$ is related to n.

vi) $7 - n$ is related to n.

10. For each relation below:

i) Describe a real-life situation that could be represented by the relation.

ii) Make a table of values.

iii) Graph the relation.

iv) Describe the graph.

v) Write 2 questions you could answer using the graph. Answer the questions.

a) $4 + 2m$ is related to m.

b) $15 - 2d$ is related to d.

11. Gerad is paid $6 to supervise a group of children at a day camp. He is paid an additional $2 per child.

a) Write a relation to show how the total amount Gerad is paid is related to the number of children supervised, c.

b) Copy and complete this table of values for the relation.

c	Amount Paid ($)
0	
5	
10	
15	

c) Draw a graph to show the relation. Describe the graph.

d) Use the graph to answer these questions:

i) How much money is Gerad paid when he supervises 25 children?

ii) Gerad was paid $46. How many children did he supervise? Show your work.

12. Suggest a real-life situation that could be represented by this graph.

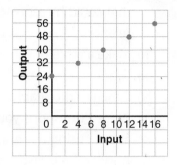

1.7 **13.** Write an equation for each sentence.

a) A pizza with 15 slices is shared equally among *n* students. Each student gets 3 slices.

b) Four less than three times the number of red counters is 20.

14. The drum ring of this hand drum is a regular octagon. It has perimeter 48 cm. Write an equation you could use to find the side length of the drum ring.

15. i) Write an equation you can use to solve each problem.

ii) Use tiles to solve each equation.

iii) Draw pictures to represent the steps you took to solve each equation.

iv) Use tiles to verify each solution.

a) Thirty-six people volunteered to canvas door-to-door for the Heart and Stroke Foundation. They were divided into groups of 3. How many groups were there?

b) A garden has 7 more daffodils than tulips. There are 18 daffodils. How many tulips are there?

c) A sleeve of juice contains 3 juice boxes. Marty buys 24 juice boxes. How many sleeves does he buy?

d) Jan collects foreign stamps. Her friend gives her 8 stamps. Jan then has 21 stamps. How many stamps did Jan have to start with?

16. A number is multiplied by 4, then 5 is added. The result is 21. What is the number?

a) Write an equation to represent this situation.

b) Solve the equation to find the number.

c) Verify the solution.

Practice Test

1. Use the digits 0 to 9.
   Replace the ☐ in 16 21☐ so that the number is divisible by:
   a) 2   **b)** 3   **c)** 4   **d)** 5
   e) 6   **f)** 8   **g)** 9   **h)** 10
   Find as many answers as you can.

2. Here are 3 algebraic expressions:
   $$2 + 3n; \qquad 2n + 3; \qquad 3n - 2$$
   Are there any values of n that will produce the same number when substituted into two or more of the expressions?
   Investigate to find out. Show your work.

3. Jamal joined a video club. The annual membership fee is $25.
   The cost of each video rental is an additional $2.

   a) Write a relation for the total cost when Jamal rents v videos in one year.
   b) Graph the relation. How much will Jamal pay when he rents 10 videos? 25 videos?
   c) How does the relation change when the cost per video rental increases by $1?
   How much more would Jamal pay to rent 10 videos?
   How do you know the answer makes sense?

4. **a)** Write an equation for each situation.
   b) Solve each equation using tiles. Sketch the tiles you used.
   c) Verify each solution.
      i) There were 22 students in a Grade 7 class.
           Five students went to a track meet.
           How many students were left in the class?
     ii) Twice the number of dogs in the park is 14.
           How many dogs are in the park?
    iii) Daphne scored the same number of goals in period one, period two, and period three this season.
           She also scored 4 overtime goals, for a total of 19 goals.
           How many goals did she score in each period?

Two students raised money for charity in a bike-a-thon.
The route was from Lethbridge to Medicine Hat, a distance of 166 km.

Part 1

Ingrid cycles 15 km each hour.
How far does Ingrid cycle in 1 h? 2 h? 3 h? 4 h? 5 h?
Record the results in a table.

Time (h)					
Distance (km)					

Graph the data.
Graph *Time* horizontally and *Distance* vertically.

Write a relation for the distance Ingrid travels in *t* hours.
How far does Ingrid travel in 7 h?
How could you check your answer?

Let *t* hours represent the time Ingrid cycled.
How far does Ingrid cycle in *t* hours?
Write an equation you can solve to find the time Ingrid took
to cycle 120 km.
Solve the equation.

Part 2

Liam cycles 20 km each hour.
Repeat *Part 1* for Liam.

Part 3

How are the graphs for Ingrid and Liam alike?
How are they different?

Part 4

Ingrid's sponsors paid her $25 per kilometre.
Liam's sponsors paid him $20 per kilometre.
Make a table to show how much money each
student raised for every 10 km cycled.

Distance (km)	Money Raised by Ingrid ($)	Money Raised by Liam ($)
10		
20		
30		

How much money did Ingrid raise if she cycled *d* kilometres?
How much money did Liam raise if he cycled *d* kilometres?

Liam and Ingrid raised equal amounts of money.
How far might each person have cycled? Explain.

Reflect on Your Learning

You have learned different ways to represent a pattern. Which way do
you find easiest to use? When might you want to use the other ways?

Integers

Canada has 6 time zones. This map shows the summer time zones.
- What time is it where you are now?
- You want to call a friend in Newfoundland. What time is it there?
- In the province or territory farthest from you, what might students be doing now?

What other questions can you ask about this map?

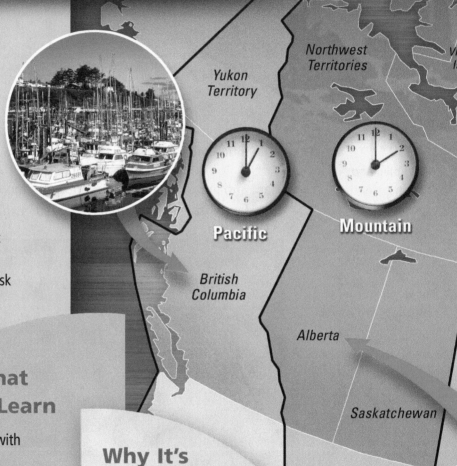

Yukon Territory

Northwest Territories

Victori Island

Pacific

Mountain

British Columbia

Alberta

Saskatchewan

What You'll Learn

- Model integers with coloured tiles.
- Add integers using coloured tiles and number lines.
- Subtract integers using coloured tiles and number lines.
- Solve problems involving the addition and subtraction of integers.

Why It's Important

- We use integers when we talk about weather, finances, sports, geography, and science.
- Integers extend the whole number work from earlier grades.

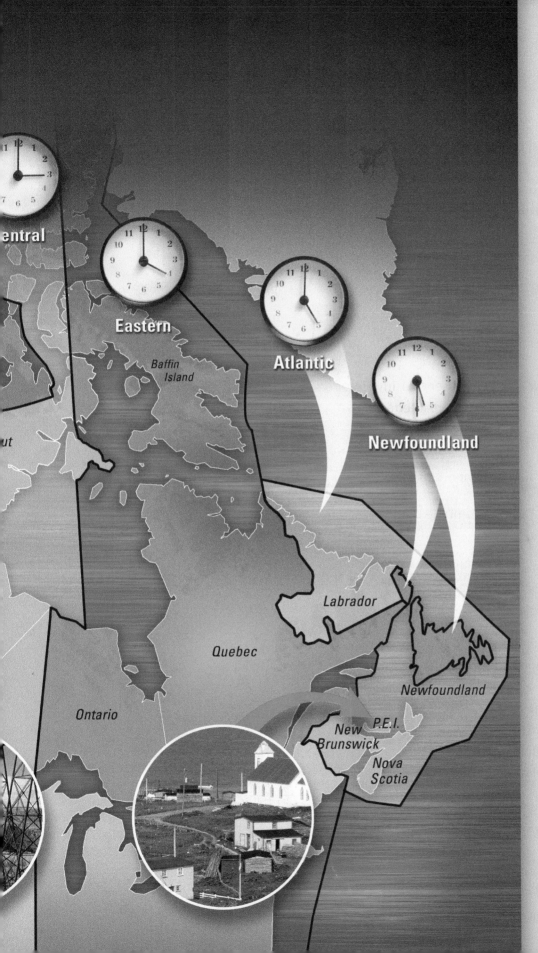

Central

Eastern

Baffin
Island

Atlantic

Newfoundland

Labrador

Quebec

Newfoundland

Ontario

New
Brunswick

P.E.I.

Nova
Scotia

Key Words

- negative integer
- positive integer
- zero pair
- opposite integers

Focus | Use coloured tiles to represent integers.

One of the coldest places on Earth is Antarctica, with an average annual temperature of about −58°C. This is a **negative integer**.

One of the hottest places on Earth is Ethiopia, with an average annual temperature of about +34°C. This is a **positive integer**.

We can use yellow tiles to represent positive integers and red tiles to represent negative integers.

One yellow tile ☐ can represent +1.

One red tile ■ can represent −1.

A red tile and a yellow tile combine to model 0:

■ −1
☐ +1

We call this a **zero pair.**

Explore

You will need coloured tiles.

➤ One of you uses 9 tiles and one uses 10 tiles. You can use any combination of red and yellow tiles each time. How many different integers can you model with 9 tiles? How many different integers can your partner model with 10 tiles?

➤ Draw a picture to show the tiles you used for each integer you modelled. Circle the zero pairs. Write the integer each picture represents. How do you know?

Reflect & Share

Compare your models with those of your partner.
Which integers did you model? Your partner?
Were you able to model any of the same integers?
Why or why not?

Connect

We can model any integer in many ways.

Each set of tiles below models +5.

-

-

Each pair of 1 yellow tile and
1 red tile makes a zero pair.
The pair models 0.

-

Example

Use coloured tiles to model −4 in three different ways.

A Solution

Start with 4 red tiles to model −4.
Add different numbers of zero pairs.
Each set of tiles below models −4.

-

Adding 4 zero pairs does not
change the value.

Adding 2 zero pairs does not
change the value.

Adding 7 zero pairs does not
change the value.

Practice

1. Write the integer modelled by each set of tiles.

a) b) c)

d) e) f)

2. Draw yellow and red tiles to model each integer in two different ways.

a) −6   b) +7   c) +6   d) −2

e) +9   f) −4   g) 0   h) +10

3. Work with a partner.

Place 10 yellow and 10 red tiles in a bag.

a) Suppose you draw 6 tiles from the bag.
   What integers might the tiles model?
   List all seven possible integers.

b) Without looking, draw 6 tiles from the bag.
   Record the integer that these tiles model.
   Repeat the experiment 9 more times.
   Which integer was modelled most often?

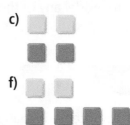

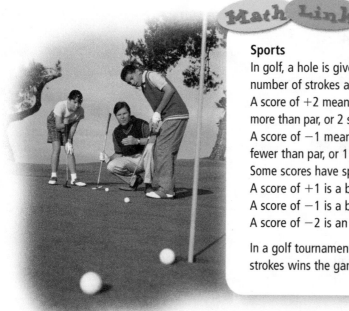

Sports

In golf, a hole is given a value called **par**. Par is the number of strokes a good golfer takes to reach the hole.
A score of +2 means a golfer took 2 strokes more than par, or 2 strokes over par.
A score of −1 means a golfer took 1 stroke fewer than par, or 1 stroke under par.
Some scores have special names.
A score of +1 is a bogey.
A score of −1 is a birdie.
A score of −2 is an eagle.

In a golf tournament, the golfer with the fewest strokes wins the game.

4. Assessment Focus

a) Choose an integer between −9 and +6.
   Use coloured tiles to model the integer.

b) How many more ways can you find to model the integer with tiles?
   Create a table to order your work.

c) What patterns can you find in your table?

d) Explain how the patterns in your table can help
   you model an integer between −90 and +60.

5. a) Suppose you have 10 yellow tiles, and use all of them.
   How many red tiles would you need to model +2?
   How do you know?

b) Suppose you have 100 yellow tiles, and use all of them.
   How many red tiles would you need to model +2?
   How do you know?

6. Write the integer suggested by each of the following situations.
   Draw yellow or red tiles to model each integer.
   Explain your choice.

a) You move your game piece forward 9 squares on
   the game board.

b) You ride down 5 floors on an elevator.

c) You walk up 11 stairs.

d) The temperature drops 9°C.

e) You climb down 7 rungs on a ladder.

7. Write two integers suggested by each of the
   following situations.

a) You deposit $100 in your bank account,
   then pay back your friend $20.

b) While shopping in a large department store,
   you ride the elevator up 6 floors, then down 4 floors.

c) The temperature rises 12°C during the day,
   then falls 8°C at night.

Reflect

How is it possible to use coloured tiles to model
any integer in many different ways?

Recall that when you add two numbers, such as 5 + 3, you can show the addition by combining 5 counters with 3 counters to obtain 8 counters.

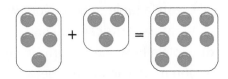

You can add two integers in a similar way.
You know that +1 and −1 combine to make a zero pair.
We can combine coloured tiles to add integers.

Explore

You will need coloured tiles.

➤ Choose two different positive integers.
   Add the integers.
   Draw a picture of the tiles you used.
   Write the addition equation.
➤ Repeat the activity for a positive integer and
   a negative integer.
➤ Repeat the activity for two different negative integers.

Reflect & Share

Share your equations with another pair of classmates.
How did you use the tiles to find a sum of integers?
How can you predict the sign of the sum?

Connect

➤ To add two positive integers: (+5) + (+4)
   We can model each integer with tiles.

   +5:

   +4: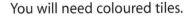

   Combine the tiles. There are 9 yellow tiles.
   They model +9.
   So, (+5) + (+4) = +9

This is an addition
equation.

➤ To add a negative integer
and a positive integer: $(-6) + (+9)$
We can model each integer with tiles. Circle zero pairs.

−6:
+9:

There are 6 zero pairs.
There are 3 yellow tiles left.
They model +3.
So, $(-6) + (+9) = +3$

➤ To add two negative integers: $(-3) + (-7)$
We can model each integer with tiles.

−3:

−7:

Combine the tiles. There are 10 red tiles.
They model −10.
So, $(-3) + (-7) = -10$

Example

The temperature rises 5°C, then falls 8°C.
a) Represent the above sentence with integers. **b)** Find the overall change in temperature.

A Solution

a) +5 represents a rise of 5°C.
−8 represents a fall of 8°C.
Using integers, the sentence is: $(+5) + (-8)$
b) Model each integer with tiles.
Circle zero pairs.

+5
−8

There are 3 red tiles left.
They model −3.
So, $(+5) + (-8) = -3$
The overall change in temperature is −3°C.

Use coloured tiles.

1. What sum does each set of tiles model?
Write the addition equation.

a)

b)

c)

d)

e)

f)

2. What sum does each set of tiles model?
How do you know you are correct?
a) 3 yellow tiles and 2 red tiles
b) 3 yellow tiles and 4 red tiles
c) 2 red tiles and 2 yellow tiles

3. Use coloured tiles to represent each sum. Find each sum.
Sketch the tiles you used. What do you notice?
a) $(+2) + (-2)$ b) $(-4) + (+4)$ c) $(+5) + (-5)$

4. Add. Sketch coloured tiles to show how you did it.
a) $(+2) + (+3)$ b) $(-3) + (+4)$ c) $(-4) + (-1)$
d) $(+1) + (-1)$ e) $(-3) + (-4)$ f) $(+5) + (-2)$

5. Add. Write the addition equations.
a) $(+4) + (+3)$ b) $(-7) + (+5)$ c) $(-4) + (-5)$
d) $(+8) + (-1)$ e) $(-10) + (-6)$ f) $(+4) + (-13)$

6. Represent each sentence with integers, then find each sum.
a) The temperature drops 3°C and rises 4°C.
b) Marie earned $5 and spent $3.
c) A stock rises 15¢, then falls 7¢.
d) Jerome moves his game piece 3 squares backward, then 8 squares forward.
e) Duma deposits $12, then withdraws $5.

7. Use question 6 as a model.

Write 3 integer addition problems.

Trade problems with a classmate.

Solve your classmate's problems with coloured tiles.

8. Copy and complete.

a) $(+5) + (\boxed{?}) = +8$

d) $(-5) + (\boxed{?}) = -3$

b) $(\boxed{?}) + (-3) = -4$

e) $(+2) + (\boxed{?}) = +1$

c) $(+3) + (\boxed{?}) = +1$

f) $(\boxed{?}) + (-6) = 0$

9. Assessment Focus

a) Add: $(+3) + (-7)$

b) Suppose you add the integers in the opposite order:
$(-7) + (+3)$. Does the sum change?
Use coloured tile drawings and words to explain the result.

c) How is $(-3) + (+7)$ different from $(+3) + (-7)$? Explain.

d) Repeat parts a to c with a sum of integers of your choice.
What do you notice?

10. Take It Further Add. Sketch coloured tiles to show how you did it.

a) $(+1) + (+2) + (+3)$

b) $(+2) + (-1) + (+3)$

c) $(-3) + (-1) + (-1)$

d) $(+4) + (-3) + (+1)$

11. Take It Further In a magic square, every row, column, and diagonal has
the same sum. Copy and complete each magic square. How did you do it?

a)

+3		+1
	0	
−1		

b)

−1		+1
	−2	
		−3

12. Take It Further Copy each integer pattern.
What do you add each time to get the next term?
Write the next 4 terms.

a) $+8, +4, 0, -4, \ldots$

b) $-12, -9, -6, -3, \ldots$

Reflect

Talk to a partner. Tell how you used coloured tiles to add
two integers when the integers have:

- the same signs
- opposite signs

2.3 Adding Integers on a Number Line

We can show the addition of whole numbers on a number line: $4 + 2 = 6$
Draw 2 arrows.

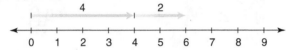

Or, begin at 4, and draw 1 arrow.

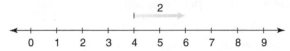

We can also show the addition of integers on a number line.

Explore

You will need copies of a number line.

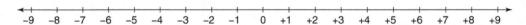

➤ Choose two different positive integers.
Use a number line to add them.
Write the addition equation.

➤ Repeat the activity for a positive integer
and a negative integer.

➤ Repeat the activity for two different
negative integers.

➤ What happens when you add $+2$ and -2?

Reflect & Share

Compare your strategies for adding on a
number line with those of your classmates.
Use coloured tiles to check the sums.
Why do you think integers such as $+2$ and -2
are called **opposite integers**?

➤ To add a positive integer, move right (in the positive direction).

(−2) + (+3)

Start at 0.

Draw an arrow 2 units long, pointing left.

This arrow represents −2.

From −2, draw an arrow 3 units long, pointing right.

This arrow represents +3.

The arrow head is at +1.

So, (−2) + (+3) = +1

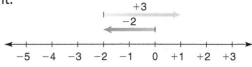

Notice that the first arrow ends at the first integer.

So, we could start at that integer,

and use only 1 arrow to find the sum.

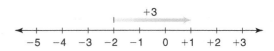

➤ To add a negative integer, move left
(in the negative direction).

(−2) + (−3)

Start at −2.

Draw an arrow 3 units long, pointing left.

This arrow represents −3.

The arrow head is at −5.

So, (−2) + (−3) = −5

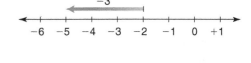

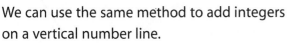

We can use the same method to add integers
on a vertical number line.

➤ The temperature is 12°C. It falls 5°C.

Find the final temperature.

(+12) + (−5)

Start at +12.

Draw an arrow 5 units long, pointing down.

This arrow represents −5.

The arrow head is at +7.

So, (+12) + (−5) = +7

The final temperature is 7°C.

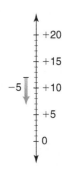

Example

Sandra and Joe buy and sell CDs at a flea market.
One day in August, they bought 3 CDs for $5 each.
They sold 2 CDs for $9 each.

a) Write the expenses and income as integers.

b) Did Sandra and Joe make money or lose money that day in August?
Explain.

A Solution

a) Expenses: $(-5) + (-5) + (-5) = -15$; they spent $15.
Income: $(+9) + (+9) = +18$; they made $18.

b) Draw a number line.
Add expenses and income.

Another Strategy
We could use coloured tiles.

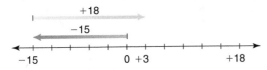

$(-15) + (+18) = +3$
Since the sum of the expenses and income is positive,
Sandra and Joe made money. They made $3.

Practice

1. Use a number line to represent each sum.

a) $(+1) + (+3)$　　b) $(-1) + (+3)$　　c) $(-3) + (+1)$　　d) $(-1) + (-3)$

e) $(-3) + (-4)$　　f) $(-3) + (+4)$　　g) $(+3) + (-4)$　　h) $(+3) + (+4)$

2. Use a number line to add.

a) $(+4) + (+2)$　　b) $(+5) + (-3)$　　c) $(-4) + (-2)$　　d) $(-8) + (+2)$

e) $(-6) + (-7)$　　f) $(+1) + (-6)$　　g) $(-5) + (+2)$　　h) $(+8) + (+4)$

3. a) Reverse the order of the integers in question 2, then add.

b) Compare your answers to the answers in question 2.
What do you notice?

c) Make a general statement about your observations.

4. Look at these thermometers. Find each temperature after:

a) it falls 4°C   **b)** it falls 7°C   **c)** it rises 6°C

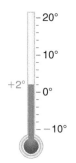

5. a) The temperature rises 7°C, then drops 2°C.
   What is the overall change in temperature?
   b) Adrian loses $4, then earns $8.
   Did Adrian gain or lose overall?
   c) The value of a stock went up $3, then down $2.
   What was the final change in the value of the stock?

6. Opposite integers are the same distance from 0
   but are on opposite sides of 0.

   a) Write the opposite of each integer.
   i) $+2$ **ii)** -5 **iii)** $+6$ **iv)** -8
   b) Add each integer to its opposite in part a.
   c) What do you notice about the sum of two
   opposite integers?

7. Use a number line. For each sentence below:
   a) Write each number as an integer.
   b) Write the addition equation.
      Explain your answer in words.
      i) You take 5 steps backward,
         then 10 steps backward.
      ii) You withdraw $5, then deposit $8.
      iii) A deep sea diver descends 8 m, then ascends 6 m.
      iv) A person drives a snowmobile 4 km east, then 7 km west.
      v) A person gains 6 kg, then loses 10 kg.

8. a) Write the addition equation modelled by each number line.

b) Describe a situation that each number line could represent.

i)

$$-5 \quad -4 \quad -3 \quad -2 \quad -1 \quad 0 \quad +1 \quad +2 \quad +3 \quad +4 \quad +5$$

ii)

$$0 \quad +1 \quad +2 \quad +3 \quad +4 \quad +5 \quad +6 \quad +7 \quad +8 \quad +9$$

9. Assessment Focus Is each statement always true, sometimes true, or never true?

Use a number line to support your answers.

a) The sum of two opposite integers is 0.

b) The sum of two positive integers is negative.

c) The sum of two negative integers is negative.

d) The sum of a negative integer and a positive integer is negative.

10. Take It Further Add.

a) $(+4) + (+3) + (-6)$

b) $(-2) + (-4) + (+1)$

c) $(-5) + (+3) + (-4)$

d) $(+6) + (-8) + (+2)$

11. Take It Further The temperature in Calgary, Alberta, was $-2°C$.
A Chinook came through and the temperature rose $15°C$.
At nightfall, it fell $7°C$. What was the final temperature?
Support your answer with a drawing.

Reflect

Compare adding on a number line to adding with coloured tiles.
Which method do you prefer?
When might you need to use a different method?

Mid-Unit Review

2.1

1. Use coloured tiles to model each integer in two different ways. Draw the tiles.
   a) -5 b) 0
   c) $+8$ d) -1
   e) $+3$ f) -7

2. Suppose you have 8 red tiles. How many yellow tiles would you need to model $+3$? How do you know?

2.2

3. What sum does each set of tiles model? How do you know you are correct? Write the addition equations.
   a) 6 yellow tiles and 1 red tile
   b) 5 yellow tiles and 7 red tiles
   c) 4 yellow tiles and 4 red tiles

4. Use coloured tiles to add. Draw pictures of the tiles you used.
   a) $(+4) + (-1)$ b) $(-3) + (-2)$
   c) $(-5) + (+1)$ d) $(+6) + (+3)$
   e) $(-4) + (-8)$ f) $(+4) + (+8)$

2.3

5. Use a number line to add. Write the addition equations.
   a) $(+3) + (+2)$ b) $(-5) + (-1)$
   c) $(-10) + (+8)$ d) $(+6) + (-5)$
   e) $(-8) + (+8)$ f) $(-5) + (+12)$

6. a) Add. $(+4) + (-5)$
   b) Find 4 different pairs of integers that have the same sum as part a.

7. Write an addition equation for each situation.
   a) Puja earned $50, and spent $20. How much did Puja then have?
   b) The temperature is 5°C, then drops 10°C. What is the final temperature?
   c) The population of a city was 124 000, then it dropped by 4000 people. What was the population then?
   d) A plane was cruising at an altitude of 12 000 m, then dropped 1200 m. What was the cruising altitude then?

8. a) Write the addition equation modelled by each number line.
   b) Describe a situation that each number line could represent.

   i)
   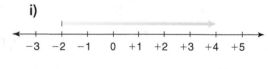

   ii)
   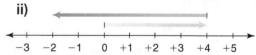

9. Each integer below is written as the sum of consecutive integers.
   $(+5) = (+2) + (+3)$
   $(+6) = (+1) + (+2) + (+3)$
   Write each of these integers as the sum of consecutive integers.
   a) $+10$ b) 0 c) $+2$
   d) $+7$ e) $+4$ f) $+8$

Focus | Use coloured tiles to subtract integers.

To add integers, we combine groups of tiles.
To subtract integers, we do the reverse:
we remove tiles from a group.

Recall that equal numbers of
red and yellow tiles model 0.
For example, +5 and −5 form 5 zero pairs,
and (−5) + (+5) = 0

Adding a zero pair to a set of tiles does not change its value.
For example, (−3) + 0 = −3

Explore

You will need coloured tiles.
Use tiles to subtract.
Add zero pairs when you need to.
Sketch the tiles you used in each case.

- (+5) − (+3)
- (+5) − (−3)
- (−3) − (+5)
- (−3) − (−5)

Reflect & Share

Compare your results with those of another pair of classmates.
Explain why you may have drawn different sets of tiles, yet both may be correct.
When you subtracted, how did you know how many tiles to use
to model each integer? How did adding zero pairs help you?

Connect

To use tiles to subtract integers, we model the first integer,
then take away the number of tiles indicated by the second integer.

We can use tiles to subtract: $(+5) - (+9)$

Model $+5$.

There are not enough tiles to take away $+9$.
To take away $+9$, we need 4 more yellow tiles.

We add zero pairs without changing the value.
Add 4 yellow tiles and 4 red tiles. They represent 0.

By adding 0, the integer the tiles represent has not changed.
Now take away the 9 yellow tiles.

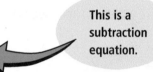

This is a subtraction equation.

Since 4 red tiles remain, we write: $(+5) - (+9) = -4$

Example

Use tiles to subtract.

a) $(-2) - (-6)$ **b)** $(-6) - (+2)$ **c)** $(+2) - (-6)$

A Solution

a) $(-2) - (-6)$

Model -2.

There are not enough tiles to take away -6.
To take away -6, we need 4 more red tiles.

We add zero pairs without changing the value.
Add 4 red tiles and 4 yellow tiles.

Now take away 6 red tiles.

Since 4 yellow tiles remain, we write: $(-2) - (-6) = +4$

b) $(-6) - (+2)$

Model -6.

There are no yellow tiles to take.
We need 2 yellow tiles to take away.

We add zero pairs.
Add 2 yellow tiles and 2 red tiles.

Now take away 2 yellow tiles.

Since 8 red tiles remain, we write: $(-6) - (+2) = -8$

c) $(+2) - (-6)$

Model $+2$.

There are no red tiles to take.
We need 6 red tiles to take away.

We add zero pairs.
Add 6 red tiles and 6 yellow tiles.

Now take away 6 red tiles.

Since 8 yellow tiles remain, we write: $(+2) - (-6) = +8$

Notice the results in the *Example*, parts b and c.
When we reverse the order in which we subtract two integers,
the answer is the opposite integer.

$(-6) - (+2) = -8$
$(+2) - (-6) = +8$

Practice

1. Use tiles to subtract. Draw pictures of the tiles you used.
   a) $(+7) - (+4)$ **b)** $(-2) - (-2)$ **c)** $(-9) - (-6)$
   d) $(+4) - (+2)$ **e)** $(-8) - (-1)$ **f)** $(+3) - (+3)$

2. Use tiles to subtract.
   a) $(-1) - (-4)$ **b)** $(+3) - (+8)$ **c)** $(-4) - (-11)$
   d) $(+7) - (+8)$ **e)** $(-4) - (-6)$ **f)** $(+1) - (+10)$

3. Subtract.
   a) $(-4) - (-1)$ **b)** $(+8) - (+3)$ **c)** $(-11) - (-4)$
   d) $(+8) - (+7)$ **e)** $(-6) + (-4)$ **f)** $(+10) - (+1)$

4. Subtract. Write the subtraction equations.
   a) $(+4) - (-7)$ **b)** $(-2) - (+8)$ **c)** $(-9) - (+5)$
   d) $(+6) - (-8)$ **e)** $(-3) - (+6)$ **f)** $(-5) - (+7)$

5. Subtract.
   a) $(+4) - (+5)$ **b)** $(-3) - (+5)$ **c)** $(-4) - (+3)$
   d) $(-1) - (-8)$ **e)** $(+8) - (-2)$ **f)** $(+4) - (-7)$

6. Use questions 1 to 5 as models.
   Write 3 integer subtraction questions.
   Trade questions with a classmate.
   Solve your classmate's questions.

7. a) Use coloured tiles to subtract each pair of integers.
   i) $(+3) - (+1)$ and $(+1) - (+3)$
   ii) $(-3) - (-2)$ and $(-2) - (-3)$
   iii) $(+4) - (-3)$ and $(-3) - (+4)$
   b) What do you notice about each pair of questions in part a?

8. $(+5) - (-2) = +7$
   Predict the value of $(-2) - (+5)$.
   Explain your prediction, then check it.

9. **Assessment Focus** Use integers.
   Write a subtraction question that would give each answer.
   How many questions can you write each time?
   a) $+2$ **b)** -3 **c)** $+5$ **d)** -6

10. Which expression in each pair has the greater value?
Explain your reasoning.
  a) i) $(+3) - (-1)$ **ii)** $(-3) - (+1)$
  b) i) $(-4) - (-5)$ **ii)** $(+4) - (+5)$

11. Take It Further
  a) Find two integers with a sum of -1 and a difference of $+5$.
  b) Create and solve a similar integer question.

12. Take It Further Copy and complete.
  a) $(+4) - \square = +3$
  b) $(+3) - \square = -1$
  c) $\square - (+1) = +4$

13. Take It Further Evaluate.
  a) $(+4) + (+1) - (+3)$
  b) $(+1) - (+2) - (-1)$
  c) $(-3) - (+1) + (+4)$
  d) $(-2) - (-4) + (-1)$
  e) $(+2) - (+1) - (+4)$
  f) $(+1) - (+2) + (+1)$

14. Take It Further Here is a magic square.
  a) Subtract $+4$ from each entry.
     Is it still a magic square? Why?
  b) Subtract -1 from each entry.
     Is it still a magic square? Why?

0	+5	−2
−1	+1	+3
+4	−3	+2

![Reflect]

Here are 4 types of subtraction questions:
  • (negative integer) − (negative integer)
  • (negative integer) − (positive integer)
  • (positive integer) − (positive integer)
  • (positive integer) − (negative integer)
Write a question for each type of subtraction.
Show how you use tiles to solve each question.

Recall how to model the subtraction of whole numbers with coloured tiles.

$7 - 5 = 2$

We can model this subtraction on a number line.

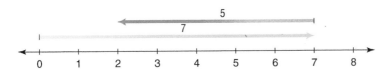

Subtraction is finding the difference. This number line shows how much more 7 is than 5.

Explore

You will need coloured tiles and copies of this number line.

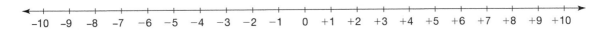

Step 1 Use tiles to subtract.
Sketch the tiles you used each time.

$(+7) - (+2)$ $(-7) - (-2)$
$(+7) - (-2)$ $(-7) - (+2)$

Step 2 Model each subtraction done with tiles on a number line.

Step 3 Use any method. Add.

$(+7) + (-2)$ $(-7) + (+2)$
$(+7) + (+2)$ $(-7) + (-2)$

Step 4 Each expression in *Step 3* has a corresponding expression in *Step 1*.
What do you notice about the answers to corresponding expressions?
What patterns do you see in each subtraction and addition?
Check your pattern using other integers.

Reflect & Share

Compare your answers with those of another pair of classmates.
How can you use addition to subtract two integers?

Connect

➤ To subtract two whole numbers, such as $5 - 2$,
we can think, "What do we add to 2 to get 5?"
We add 3 to 2 to get 5; so, $5 - 2 = 3$

We could also think:
How much more is
5 than 2?

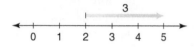

➤ We can do the same to subtract two integers.
For example, to subtract: $(+5) - (-2)$
Think: "What do we add to -2 to get $+5$?"

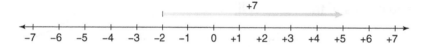

We add $+7$ to -2 to get $+5$; so, $(+5) - (-2) = +7$
We also know that $(+5) + (+2) = +7$.
We can look at other subtraction equations and
related addition equations.

$(+9) - (+4) = +5$ $(+9) + (-4) = +5$
$(-9) - (-4) = -5$ $(-9) + (+4) = -5$
$(-9) - (+4) = -13$ $(-9) + (-4) = -13$
$(+9) - (-4) = +13$ $(+9) + (+4) = +13$

In each case, the result of subtracting an integer is the
same as adding the opposite integer.
For example,

$(-9) - (+4) = -13$ $(-9) + (-4) = -13$

Subtract $+4$.   Add -4.

➤ To subtract an integer, we add the opposite integer.
For example, to subtract: $(-3) - (-6)$
Add the opposite: $(-3) + (+6)$

The opposite of -6 is $+6$.

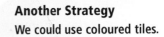

So, $(-3) - (-6) = +3$

Example

Subtract.

a) $(+2) - (+9)$ **b)** $(-2) - (+9)$

A Solution

a) To subtract: $(+2) - (+9)$
Add the opposite: $(+2) + (-9)$
Use a number line.
$(+2) + (-9) = -7$

Another Strategy
We could use coloured tiles.

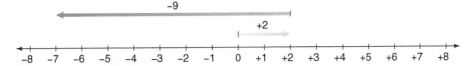

b) To subtract: $(-2) - (+9)$
Add the opposite: $(-2) + (-9)$
Use a number line.
$(-2) + (-9) = -11$

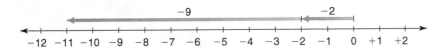

Practice

1. Use a number line to subtract.
Use coloured tiles to check your answers.

a) $(+2) - (+1)$ **b)** $(+4) - (-3)$ **c)** $(-4) - (-1)$
d) $(-5) - (+2)$ **e)** $(-2) - (-6)$ **f)** $(-3) - (-7)$

2. a) Reverse the order of the integers in question 1, then subtract.
   b) How are the answers different from those in question 1? Explain.

3. Use a number line to subtract. Write the subtraction equations.
   a) $(+10) - (+5)$ **b)** $(+7) - (-3)$ **c)** $(-8) - (+6)$
   d) $(-10) - (+5)$ **e)** $(-4) - (+4)$ **f)** $(-4) - (-4)$

4. Rewrite using addition to find each difference.
   a) $(+6) - (+4)$ **b)** $(-5) - (+4)$ **c)** $(-2) - (-3)$
   d) $(+4) - (-2)$ **e)** $(+1) - (+1)$ **f)** $(+1) - (-1)$

5. What is the difference in temperatures?
   How can you subtract to find out?
   a) A temperature 7°C above zero and a temperature 5°C below zero
   b) A temperature 15°C below zero and a temperature 8°C below zero
   c) A temperature 4°C below zero and a temperature 9°C above zero

6. What is the difference in golf scores?
   How can you subtract to find out?
   a) A golf score of 2 over par and a golf score of 6 under par
   b) A golf score of 3 under par and a golf score of 8 under par
   c) A golf score of 5 under par and a golf score of 4 over par

7. a) The table shows the average afternoon
   temperatures in January and April for four
   Canadian cities.
   What is the rise in temperature from January
   to April for each city? Show your work.
   b) Which city has the greatest difference
   in temperatures?
   How do you know?

	City	January Temperature	April Temperature
i)	Calgary	−4°C	+13°C
ii)	Iqaluit	−22°C	−10°C
iii)	Toronto	−3°C	+12°C
iv)	Victoria	+7°C	+13°C

8. **Assessment Focus**
   a) Subtract: $(-6) - (+11)$
   b) Suppose we subtract the integers in the opposite order: $(+11) - (-6)$
      How does the answer compare with the answer in part a?
      Use number lines to explain.
   c) How is $(+6) - (-11)$ different from $(-6) - (+11)$? Explain.

9. Show three ways that $+4$ can be written as the difference of two integers.

10. **Take It Further** Use patterns to subtract.
    a) Subtract: $(+2) - (+5)$
       Start the pattern with $(+6) - (+5) = +1$.
    b) Subtract: $(+7) - (-3)$
       Start the pattern with $(+7) - (+4) = +3$.
    c) Subtract: $(-3) - (+7)$
       Start the pattern with $(+8) - (+7) = +1$.

11. **Take It Further** Copy each integer pattern.
    Write the next 4 terms.
    What is the pattern rule?
    a) $+6, +2, -2, \dots$ b) $-3, -1, +1, \dots$
    c) $+5, +12, +19, \dots$ d) $+1, 0, -1, \dots$

12. **Take It Further** Evaluate.
    a) $(+4) - (+2) - (+1)$ b) $(-2) - (+1) - (-4)$
    c) $(-1) + (-2) - (+1)$ d) $(+5) - (+1) + (-2)$
    e) $(+10) - (+3) - (-5)$ f) $(-7) - (+1) + (-3)$

Reflect

How is the subtraction of integers related to the addition of integers?
Use coloured tiles or a number line to show your thinking.

Writing to Reflect on Your Understanding

As you work through a math unit, you will come across many new ideas.

Sometimes it is hard to decide what you already know.
What you know can often help you understand the new ideas.

You can use a Homework Log to help you reflect on your understanding.

Using a Homework Log

As you work through your homework, ask yourself:
- What is the key idea?
- How difficult is the homework for me?
- Which questions am I able to do?
- Which questions do I need help with?
- What questions could I ask to help me with my homework?

Tips for Writing a Homework Log

- Write so that someone else can understand you.
- Write out a question that you cannot solve.
- Describe 3 ways you tried to solve the question.
- Write a question you can ask to help you better understand your homework.

Here is a sample Homework Log.

Name: Asad

Homework Log

The homework was . . .   P26 #5-12

The key concept was . . . Subtracting integers on a number line

Overall, I'd rate the difficulty level of the homework as . . .

Easy $\longmapsto$ 0 1 2 3 4 5 6 7 8 9 10 Hard

One question I had difficulty with but solved was . . .
What is the difference in temperatures?
A temperature 7°C above zero and a temperature 5°C below zero

7°C
0°C } 12°C
-5°C

A question I couldn't solve was . . .
Evaluate: $(+5) - (+1) + (-2)$

To solve it I tried these things . . .
1. I used my calculator but I know I should be able to do it without one.
2. I tried to model it on a number line but I didn't know which way to draw the arrows.
3. I looked at the example in the book. It says to "add the opposite" but I don't know what that means.

Questions for experts . . .
What does "add the opposite" mean?
How do you take away a negative integer?

✓ Check

- Complete a Homework Log for your next homework assignment.
- Share your Homework Log with a classmate.
- Try to help each other with questions that you were unable to solve.

Unit Review

What Do I Need to Know?

☑ **Adding Integers**
- You can use tiles to add integers.
  $(-7) + (+2) = -5$

- You can use a number line to add integers.
  $(+6) + (-3) = +3$

☑ **Subtracting Integers**
- You can use tiles to subtract integers: $(+3) - (-7)$
  We need enough red tiles to take away 7 of them.
  Model $+3$:

  Since there are not enough tiles to take away -7, add 7 yellow tiles and 7 red tiles. Now take away 7 red tiles. There are 10 yellow tiles left.

  $(+3) - (-7) = +10$

- You can also subtract by adding the opposite:
  $(-5) - (-8) = (-5) + (+8)$
  $\qquad\qquad\quad = +3$

- You can use a number line to subtract integers.
  $(-4) - (+7)$
  Add the opposite: $(-4) + (-7)$
  Use a number line.
  $(-4) - (+7) = -11$

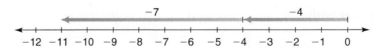

What Should I Be Able to Do?

LESSON

2.1 **1.** Suppose you have 17 red tiles.
How many yellow tiles would you
need to model:
a) −12?   b) 0?
c) +20?   d) −17?
How do you know?

2. Write the integer suggested by each
of the following situations.
Draw yellow or red tiles to model
each integer.
Explain your choice.
a) The temperature rises 8°C.
b) The price of 1 L of gas falls 5¢.
c) You deposit $12 in your
bank account.
d) You take 7 steps backward.
e) The time is 9 s before take-off.

2.2 **3.** What sum does each set of
tiles model?
a) 5 red tiles and 2 yellow tiles
b) 6 yellow tiles and 5 red tiles
c) 6 yellow tiles and 7 red tiles
d) 8 yellow tiles and 8 red tiles

4. Represent each sentence with
integers, then find each sum.
a) The temperature was −6°C,
then rose 4°C.
b) Surinder withdrew $25,
then deposited $13.
c) A stock gained $15,
then lost $23.
d) A submarine was 250 m below
sea level, then ascended 80 m.

5. a) Find 4 pairs of integers that have
the sum −5.
b) Find 4 pairs of integers that have
the sum +4.

2.3 **6.** The temperature at 6 a.m. is −10°C.
During the day, the temperature
rises 17°C. What is the new
temperature? Write an addition
equation to represent this situation.
Use a vertical number line to
support your answer.

7. a) Write an addition equation
modelled by each number line.
b) Describe a situation that each
number line could represent.
i)

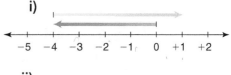

ii)

2.2 **8.** Use tiles to add or subtract.
2.4
a) (−1) + (+3)
b) (+3) + (−4)
c) (−2) − (+3)
d) (−1) − (−3)

2.3
2.5

9. Use a number line to add or subtract.
   a) $(-1) + (+3)$ b) $(+6) + (-4)$
   c) $(-4) - (+6)$ d) $(-5) - (-3)$

10. When you add two positive integers, their sum is always a positive integer.
   When you subtract two positive integers, is their difference always a positive integer? Explain.

11. a) What temperature is 7°C warmer than 2°C?
   b) What temperature is 5°C warmer than −5°C?
   c) What temperature is 8°C cooler than 2°C?
   d) What temperature is 4°C cooler than −3°C?

2.4
2.5

12. Use tiles or a number line to subtract. Write the subtraction equations.
   a) $(+4) - (+1)$ b) $(+5) - (-1)$
   c) $(+2) - (-2)$ d) $(-4) - (+1)$
   e) $(-6) - (-2)$ f) $(-10) - (-5)$
   g) $(-4) - (-2)$ h) $(-5) - (-10)$

13. Subtract.
   a) $(+7) - (+2)$ b) $(-7) - (+3)$
   c) $(-4) - (-5)$ d) $(+3) - (+3)$
   e) $(+3) - (-3)$ f) $(-3) - (-2)$

14. Use tiles or a number line.
   Find the difference between:
   a) a temperature of +5°C and −7°C
   b) an elevation of −100 m and +50 m

15. What is the difference in heights? How can you subtract to find out?
   a) A water level of 2 m below sea level and a water level of 7 m above sea level
   b) A balloon 25 m above ground and a balloon 11 m above ground

16. What is the difference in masses? How can you subtract to find out?
   a) A gain of 9 kg and a loss of 3 kg
   b) A loss of 6 kg and a loss of 5 kg

17. We measure time in hours. Suppose 12 noon is represented by the integer 0.
   a) Which integer represents 1 p.m. the same day?
   b) Which integer represents 10 a.m. the same day?
   c) Find the difference between these times in 2 ways. Show your work.

18. a) Find 5 pairs of integers with a difference of +6.
   b) Find 5 pairs of integers with a difference of −3.

Practice Test

1. Evaluate. Use coloured tiles.
   Record your work.
   a) $(+5) + (-8)$ b) $(-3) - (+7)$ c) $(-9) + (-1)$
   d) $(-4) + (+10)$ e) $(-6) - (-2)$ f) $(+12) - (-11)$

2. Evaluate. Use a number line.
   Record your work.
   a) $(+9) + (-1)$ b) $(-4) - (+11)$ c) $(-8) + (-3)$
   d) $(+13) - (+6)$ e) $(-7) + (+9)$ f) $(-1) - (-5)$

3. Without calculating the sum, how can you tell if the
   sum of two integers will be:
   a) zero?   b) negative?   c) positive?
   Include examples in your explanations.

4. Here is a different type of dartboard.
   A player throws 2 darts at the board.
   His score is the sum of the integers
   in the areas his darts land.
   Assume both darts hit the board.
   a) How many different scores are possible?
   b) Find each score.

5. The lowest temperature possible is approximately $-273°C$.
   The temperature at which water boils is $100°C$.
   What is the difference in these temperatures?

6. Place 3 integers in a row as shown.

   $(+6)$ $(+4)$ (-3)
   How many different answers can you get by putting addition
   and/or subtraction signs between the integers?
   How do you know you have found all possible answers?
   For example: $(+6) + (+4) - (-3)$
   What if there were 4 integers in a row?

TIME ZONES

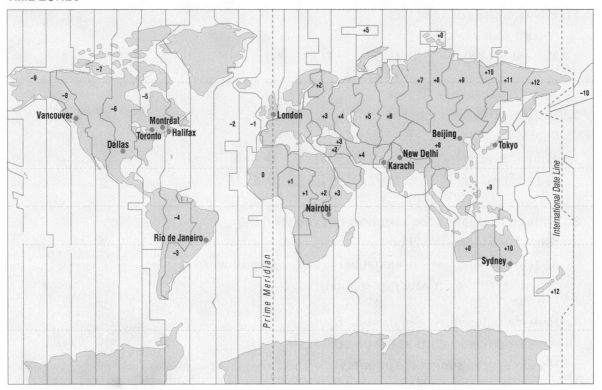

The map shows the world's time zones.

Greenwich, in London, England, is the reference point, or the zero for the time zones. Its time is called UTC, or Coordinated Universal Time. London, England, is also in this time zone.

The positive and negative integers on the map show the changes in time from UTC.

The 2008 Summer Olympics will be held in Beijing, China.

1. The local start times of some Olympic events are given. Family members want to watch these events live, in Brandon (the same time zone as Dallas).
   What time should they "tune in"? How do you know?
   a) 200-m backstroke at 2:00 p.m.
   b) 100-m dash at 7:00 p.m.
   c) gymnastics at 11:00 p.m.
   d) middleweight boxing at 8:00 a.m.

2. An event is broadcast live in Montreal at 9:00 p.m.
   What time is it taking place in Beijing?
   Show your work.

3. Two pen pals plan to meet in Beijing for the Olympics.
   Atsuko lives in Tokyo, Japan.
   She can get a direct flight to Beijing that takes 4 h.
   Paula lives in Sydney, Australia, and her direct flight takes 13 h.
   What time does each girl leave her country to arrive in Beijing
   at 6 p.m., Beijing time?

Check List

Your work should show:

✓ how you used integers to solve each problem

✓ the strategies you used to add and subtract integers

✓ correct calculations

✓ a clear explanation of how integers relate to time zones

4. Olympic funding depends on money from North American
   television networks. What problems will the organizers of
   the Beijing Olympics encounter when they plan the times
   for events?

5. Make up your own problem about the time zone map.
   Solve your problem. Show your work.

Show how you can use integers to solve each problem.

Reflect on Your Learning

Suppose there were no negative integers.
Could we survive in a world without negative integers?
Explain.

UNIT 3

Fractions, Decimals, and Percents

Ice Skates
$50.00

Skis
$200.00

Running
Shoes
$40.00

Stores offer goods on sale to encourage
you to spend money.
Look at these advertisements.
What is the sale price of each item
in the picture using each advertisement?
How did you calculate the sale price?
Explain your strategy.

What You'll Learn

- Convert between fractions
  and terminating or repeating
  decimals.
- Compare and order fractions,
  decimals, and mixed numbers.
- Add, subtract, multiply, and
  divide decimals.
- Solve problems involving fractions,
  decimals, and percents.

Why It's Important

- You use fractions and decimals
  when you shop, measure,
  and work with a percent;
  and in sports, recipes, and business.
- You use percents when you shop, to
  find sale prices and to calculate taxes.

Hockey Sweater $80.00

EVERYTHING 25% OFF

All items $\frac{1}{2}$ off

Key Words

- terminating decimal
- repeating decimal

3.1 Fractions to Decimals

Focus Use patterns to convert between decimals and fractions.

Numbers can be written in both fraction and decimal form.

For example, 3 can be written as $\frac{3}{1}$ and 3.0.

A fraction illustrates division;

that is, $\frac{1}{10}$ means $1 \div 10$.

Recall that $\frac{1}{10}$ is 0.1 in decimal form.

$\frac{3}{100}$ is 0.03 in decimal form.

$\frac{45}{1000}$ is 0.045 in decimal form.

Here are some more fractions and decimals you learned in earlier grades.

Fraction	$\frac{7}{10}$	$\frac{1}{100}$	$\frac{19}{100}$	$\frac{1}{1000}$	$\frac{23}{1000}$	$\frac{471}{1000}$
Decimal	0.7	0.01	0.19	0.001	0.023	0.471

Explore

You will need a calculator.

➤ Use a calculator.

Write each fraction as a decimal: $\frac{1}{11}, \frac{2}{11}, \frac{3}{11}, \frac{4}{11}$

What patterns do you see?

Use your patterns to predict the decimal forms of these fractions:

$\frac{5}{11}, \frac{6}{11}, \frac{7}{11}, \frac{8}{11}, \frac{9}{11}, \frac{10}{11}$

Use a calculator to check your predictions.

➤ Use a calculator.

Write each fraction as a decimal: $\frac{1}{9}, \frac{2}{9}, \frac{3}{9}$

What patterns do you see?

Use your patterns to predict the fraction form of these decimals:

0.777 777 777...   0.888 888 888...

Check your predictions.

What do you notice about the last digit in the calculator display?

Reflect & Share

Compare your patterns, decimals, and fractions with those of another pair of classmates. How did you use patterns to make predictions?

Connect

➤ Decimals, such as 0.1 and 0.25, are **terminating decimals**.
Each decimal has a definite number of decimal places.

➤ Decimals, such as 0.333 333 333…; 0.454 545 454…; 0.811 111 111…
are **repeating decimals**.
Some digits in each repeating decimal repeat forever.
We draw a bar over the digits that repeat.
For example, $\frac{4}{33} = 4 \div 33 = 0.121\ 212\ 121…$, which is written as $0.\overline{12}$
$\frac{73}{90} = 73 \div 90 = 0.811\ 111\ 111…$, which is written as $0.81\overline{1}$

➤ Patterns sometimes occur when we write fractions in decimal form.
For example,
$\frac{1}{99} = 0.\overline{01}$ $\frac{2}{99} = 0.\overline{02}$ $\frac{15}{99} = 0.\overline{15}$ $\frac{43}{99} = 0.\overline{43}$
For fractions with denominator 99, the digits in the numerator
of the fraction are the repeating digits in the decimal.
We can use this pattern to make predictions.
To write $0.\overline{67}$ as a fraction, write the repeating digits, 67,
as the numerator of a fraction with denominator 99.
$0.\overline{67} = \frac{67}{99}$
Similarly, $0.\overline{7} = 0.\overline{77} = \frac{77}{99} = \frac{7}{9}$

Example

a) Write each fraction as a decimal.
b) Sort the fractions as representing repeating or terminating decimals:
   $\frac{13}{200}, \frac{1}{5}, \frac{11}{20}, \frac{3}{7}$

A Solution

a) Try to write each fraction with denominator 10, 100, or 1000.

$$\frac{13}{200} \overset{\times 5}{\underset{\times 5}{=}} \frac{65}{1000}, \text{ or } 0.065$$

$$\frac{1}{5} \overset{\times 2}{\underset{\times 2}{=}} \frac{2}{10}, \text{ or } 0.2$$

$$\frac{11}{20} = \frac{55}{100}, \text{ or } 0.55$$

(×5 shown over and under the equation)

$\frac{3}{7}$ cannot be written as a fraction with denominator 10, 100, or 1000.

Use a calculator.

$\frac{3}{7} = 3 \div 7 = 0.428\ 571\ 429$

This appears to be a terminating decimal.

We use long division to check.

Since we are dividing by 7, the remainders must be

less than 7.

Since we get a remainder that occurred before,

the division repeats.

So, $\frac{3}{7} = 0.\overline{428\ 571}$

The calculator rounds the decimal to fit the display:

$\frac{3}{7} = 0.428\ 571\ 428\ 571\ldots$

This is the last digit in the display.   **Since this digit is 5, the calculator adds 1 to the preceding digit.**

```
       0.4285714
7) 3.00000000
   28
    20
    14
    60
    56
    40
    35
    50
    49
    10
     7
    30
    28
    20
```

So, the calculator displays an approximate decimal value:

$\frac{3}{7} \doteq 0.428\ 571\ 429$

b) Since 0.065, 0.2, and 0.55 terminate, $\frac{13}{200}$, $\frac{1}{5}$, and $\frac{11}{20}$ represent terminating decimals.
Since $0.\overline{428\ 571}$ repeats, $\frac{3}{7}$ represents a repeating decimal.

Practice

Use a calculator when you need to.

1. a) Write each fraction as a decimal.

    i) $\frac{2}{3}$ **ii)** $\frac{3}{4}$ **iii)** $\frac{4}{5}$ **iv)** $\frac{5}{6}$ **v)** $\frac{6}{7}$

  b) Identify each decimal as terminating or repeating.

2. Write each decimal as a fraction.

    a) 0.9   **b)** 0.26   **c)** 0.45   **d)** 0.01   **e)** 0.125

3. a) Write each fraction as a decimal.

i) $\frac{1}{27}$ 　　　　　 ii) $\frac{2}{27}$ 　　　　　 iii) $\frac{3}{27}$

b) Describe the pattern in your answers to part a.

c) Use your pattern to predict the decimal form of each fraction.

i) $\frac{4}{27}$ 　　　　　 ii) $\frac{5}{27}$ 　　　　　 iii) $\frac{8}{27}$

4. For each fraction, write an equivalent fraction with denominator 10, 100, or 1000. Then, write the fraction as a decimal.

a) $\frac{2}{5}$ 　　　 **b)** $\frac{1}{4}$ 　　　 **c)** $\frac{13}{25}$ 　　　 **d)** $\frac{19}{50}$ 　　　 **e)** $\frac{37}{500}$

5. Write each decimal as a fraction in simplest form.

a) $0.\overline{6}$ 　　　 **b)** $0.\overline{5}$ 　　　 **c)** $0.\overline{41}$ 　　　 **d)** $0.1\overline{6}$

6. Write each fraction as a decimal.

a) $\frac{4}{7}$ 　　　 **b)** $\frac{4}{9}$ 　　　 **c)** $\frac{6}{11}$ 　　　 **d)** $\frac{7}{13}$

7. Write $\frac{5}{17}$ as a decimal.

The calculator display is not long enough to show the repeating digits.
How could you find the repeating digits?

8. Write $\frac{1}{5}$ as a decimal.

Use this decimal to write each number below as a decimal.

a) $\frac{4}{5}$ 　　　 **b)** $\frac{7}{5}$ 　　　 **c)** $\frac{9}{5}$ 　　　 **d)** $\frac{11}{5}$

9. a) Write each fraction as a decimal.

i) $\frac{1}{999}$ 　　　 ii) $\frac{2}{999}$ 　　　 iii) $\frac{54}{999}$ 　　　 iv) $\frac{113}{999}$

b) Describe the pattern in your answers to part a.

c) Use your pattern to predict the fraction form of each decimal.

i) $0.\overline{004}$ 　　　 ii) $0.\overline{089}$ 　　　 iii) $0.\overline{201}$ 　　　 iv) $0.\overline{326}$

10. Match each set of decimals and fractions.
Explain how you know.

a) $\frac{1}{3}, \frac{2}{3}, \frac{3}{3}, \frac{4}{3}, \frac{5}{3}$ 　　　　　 i) $0.125, 0.25, 0.375, 0.5, 0.625$

b) $\frac{1}{8}, \frac{2}{8}, \frac{3}{8}, \frac{4}{8}, \frac{5}{8}$ 　　　　　 ii) $0.1\overline{6}, 0.\overline{3}, 0.5, 0.\overline{6}, 0.8\overline{3}$

c) $\frac{1}{5}, \frac{2}{5}, \frac{3}{5}, \frac{4}{5}, \frac{5}{5}$ 　　　　　 iii) $0.\overline{3}, 0.\overline{6}, 1.0, 1.\overline{3}, 1.\overline{6}$

d) $\frac{1}{6}, \frac{2}{6}, \frac{3}{6}, \frac{4}{6}, \frac{5}{6}$ 　　　　　 iv) $0.2, 0.4, 0.6, 0.8, 1.0$

11. Assessment Focus Here is the Fibonacci sequence:

1, 1, 2, 3, 5, 8, 13, 21, 34, 55, 89, …

We can write consecutive terms as fractions:

$\frac{1}{1}, \frac{2}{1}, \frac{3}{2}, \frac{5}{3}, \frac{8}{5}, \frac{13}{8}$, and so on

   a) Write each fraction above as a decimal.
   What do you notice about the trend in the decimals?

   b) Continue to write consecutive terms as decimals.
   Write about what you find out.

12. a) Write $\frac{1}{7}$ as a repeating decimal.
   How many digits repeat?
   These repeating digits are shown around the circle
   at the right.

   b) Write the fractions $\frac{2}{7}, \frac{3}{7}, \frac{4}{7}, \frac{5}{7}$, and $\frac{6}{7}$ in decimal form.
   What patterns do you see?
   Explain how the circle of digits can help you write
   these fractions as decimals.

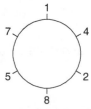

13. Take It Further

   a) Write each fraction as a decimal.
   Identify the decimals as repeating or terminating.

   i) $\frac{7}{8}$ **ii)** $\frac{5}{18}$ **iii)** $\frac{3}{10}$ **iv)** $\frac{8}{27}$ **v)** $\frac{4}{25}$

   b) Write the denominator of each fraction in part a
   as a product of prime factors.

   c) What do you notice about the prime factors
   of the denominators of the terminating decimals?
   The repeating decimals?

   d) Use your answers to part c.
   Predict which of these fractions can be written
   as terminating decimals.

   i) $\frac{7}{15}$ **ii)** $\frac{13}{40}$ **iii)** $\frac{5}{81}$ **iv)** $\frac{9}{16}$

> A prime number has
> exactly two factors,
> itself and 1. We can
> write 12 as a product
> of prime factors:
> 2 × 2 × 3

Reflect

Sometimes it is hard to figure out if a fraction can be written
as a terminating decimal or a repeating decimal.
What can you do if you are stuck?

Comparing and Ordering Fractions and Decimals

Focus Use benchmarks, place value, and equivalent fractions to compare and order fractions and decimals.

Recall how to use the benchmarks $0, \frac{1}{2}$, and 1 to compare fractions.

For example, $\frac{3}{20}$ is close to 0 because the numerator is much less than the denominator.

$\frac{11}{20}$ is close to $\frac{1}{2}$ because the numerator is about $\frac{1}{2}$ the denominator.

$\frac{19}{20}$ is close to 1 because the numerator and denominator are close in value.

Explore

Use any materials to help.

Dusan, Sasha, and Kimberley sold chocolate bars as a fund-raiser for their choir.

The bars were packaged in cartons, but sold individually.

Dusan sold $2\frac{2}{3}$ cartons. Sasha sold $\frac{5}{2}$ cartons. Kimberley sold 2.25 cartons.

Who sold the most chocolate bars?

Reflect & Share

Share your solution with another pair of classmates.

How did you decide which number was greatest?

Did you use any materials to help? How did they help?

Try to find a way to compare the numbers without using materials.

Connect

Any fraction greater than 1 can be written as a mixed number.

The benchmarks $0, \frac{1}{2}$, and 1 can be used to compare the fraction parts of mixed numbers.

We can use benchmarks on a number line to order these numbers: $\frac{2}{11}, 2\frac{3}{8}, 1\frac{1}{16}, \frac{14}{9}, \frac{14}{15}$

$\frac{2}{11}$ is close to 0.

Since $\frac{3}{8}$ is close to $\frac{1}{2}$, but less than $\frac{1}{2}$,

$2\frac{3}{8}$ is close to $2\frac{1}{2}$, but less than $2\frac{1}{2}$.

$1\frac{1}{16}$ is close to 1, but greater than 1.

$$\frac{14}{9} = \frac{9}{9} + \frac{5}{9} = 1\frac{5}{9}$$

$1\frac{5}{9}$ is close to $1\frac{1}{2}$, but greater than $1\frac{1}{2}$.

$\frac{14}{15}$ is close to 1, but less than 1.

Place the fractions on a number line.

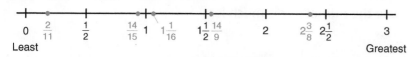

Least   Greatest

The numbers in order from greatest to least are: $2\frac{3}{8}, \frac{14}{9}, 1\frac{1}{16}, \frac{14}{15}, \frac{2}{11}$

We can also use equivalent fractions to order fractions.

Example

a) Write these numbers in order from least to greatest: $\frac{7}{8}, \frac{9}{8}, 1\frac{1}{4}, 0.75$

b) Write a fraction between $\frac{9}{8}$ and $1\frac{1}{4}$.

A Solution

a) Write equivalent fractions with like denominators, then compare the numerators.

First write the decimal as a fraction: $0.75 = \frac{75}{100} = \frac{3}{4}$

Compare: $\frac{7}{8}, \frac{9}{8}, 1\frac{1}{4}, \frac{3}{4}$

Since 8 is a multiple of 4, use 8 as a common denominator.

$$1\frac{1}{4} = \frac{4}{4} + \frac{1}{4}$$
$$= \frac{5}{4}$$

$$\frac{3}{4} \xrightarrow{\times 2} \frac{6}{8} \quad (\times 2)$$

$$\frac{5}{4} \xrightarrow{\times 2} \frac{10}{8} \quad (\times 2)$$

Each fraction now has denominator 8: $\frac{7}{8}, \frac{9}{8}, \frac{10}{8}, \frac{6}{8}$

Compare the numerators: $6 < 7 < 9 < 10$

So, $\frac{6}{8} < \frac{7}{8} < \frac{9}{8} < \frac{10}{8}$

So, $0.75 < \frac{7}{8} < \frac{9}{8} < 1\frac{1}{4}$

We can verify this order by placing the numbers on a number line.

b) Use the equivalent fraction for $1\frac{1}{4}$ with denominator 8 from part a: $\frac{10}{8}$

Find a fraction between $\frac{9}{8}$ and $\frac{10}{8}$.

The numerators are consecutive whole numbers. There are no whole numbers between 9 and 10. Multiply the numerator and denominator of both fractions by the same number to get equivalent fractions.

Choose 2:

$$\frac{9}{8} = \frac{18}{16} \qquad\qquad \frac{10}{8} = \frac{20}{16}$$

We can verify this using a number line.

Look at the numerators.

19 is between 18 and 20,

so $\frac{19}{16}$ is between $\frac{18}{16}$ and $\frac{20}{16}$.

So, $\frac{19}{16}$, or $1\frac{3}{16}$, is between $\frac{9}{8}$ and $1\frac{1}{4}$.

Another Solution

We can also use place value to order decimals.

a) Write each number as a decimal.

$$\frac{7}{8} = 0.875 \qquad \frac{9}{8} = 1.125 \qquad 1\frac{1}{4} = 1.25 \qquad 0.75$$

Write each decimal in a place-value chart.
Compare the ones.
Two numbers have 1 one and two numbers have 0 ones.

Ones	Tenths	Hundredths	Thousandths
0	8	7	5
1	1	2	5
1	2	5	0
0	7	5	0

Look at the decimals with 0 ones: 0.**8**75, 0.**7**50
Compare the tenths: 7 tenths is less than 8 tenths, so 0.750 < 0.875

Look at the decimals with 1 one: 1.**1**25 and 1.**2**50
Compare the tenths: 1 tenth is less than 2 tenths, so 1.125 < 1.250

The numbers in order from least to greatest are: 0.750, 0.875, 1.125, 1.250
So, $0.75 < \frac{7}{8} < \frac{9}{8} < 1\frac{1}{4}$

We can verify this using a number line.

b) $\frac{9}{8} = 1.125$ $\qquad$ $1\frac{1}{4} = 1.25$

Use the number line above.

1.2 lies between 1.125 and 1.25.

Write 1.2 as a fraction.

1.2 is $1\frac{2}{10}$, or $1\frac{1}{5}$.

So, $1\frac{1}{5}$, or $\frac{6}{5}$, lies between $\frac{9}{8}$ and $1\frac{1}{4}$.

There are many other possible fractions between $\frac{9}{8}$ and $1\frac{1}{4}$.

Practice

1. Write 5 different fractions with like denominators.
Draw a number line, then order the fractions on the line.
Explain your strategy.

2. Use 1-cm grid paper.
Draw a 12-cm number line like the one shown.
Use the number line to order these
numbers from greatest to least.
$2\frac{1}{2}, \frac{11}{3}, 2\frac{5}{6}$

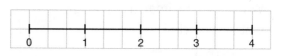

3. Use benchmarks and a number line to order each set of numbers
from least to greatest.
a) $\frac{7}{6}, \frac{15}{12}, 1\frac{2}{9}, 1$ $\qquad$ b) $1\frac{3}{4}, \frac{7}{3}, \frac{7}{6}, 2$ $\qquad$ c) $\frac{7}{4}, \frac{15}{10}, \frac{11}{5}, 2$ $\qquad$ d) $\frac{10}{4}, 2\frac{1}{3}, \frac{9}{2}, 3$

4. Use equivalent fractions.
Order each set of numbers from greatest to least.
Verify by writing each fraction as a decimal.
a) $3\frac{1}{2}, \frac{13}{4}, 3\frac{1}{8}$ $\qquad$ b) $\frac{5}{6}, \frac{2}{3}, 1\frac{1}{12}, \frac{9}{12}$ $\qquad$ c) $1\frac{2}{5}, \frac{4}{3}, \frac{3}{2}$

5. Use place value.
Order each set of numbers from least to greatest.
Verify by using a number line.
a) $\frac{7}{4}, 1.6, 1\frac{4}{5}, 1.25, 1$ $\qquad$ b) $2\frac{5}{8}, 1.875, 2\frac{3}{4}, \frac{5}{2}, 2$

6. a) Use any method. Order these numbers from greatest to least. Explain the method you used.

$\frac{17}{5}, 3.2, 2.8, 3\frac{1}{4}, \frac{21}{7}, 2$

b) Use a different method. Verify your answer in part a.

7. Find a number between the two numbers represented by each pair of dots.

a)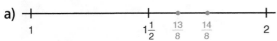

b)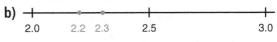

8. Find a number between each pair of numbers.

a) $\frac{5}{7}, \frac{6}{7}$ 　　　　**b)** $1\frac{2}{5}, \frac{8}{5}$ 　　　　**c)** $1.3, 1\frac{2}{5}$ 　　　　**d)** $0.5, 0.6$

9. Identify the number that has been placed incorrectly. Explain how you know.

a)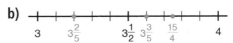

b)

10. In each set, identify the number that is not in the correct order. Show where it should go. Explain your work.

a) $\frac{29}{5}, 6\frac{2}{10}, 6.25, 6\frac{2}{20}$ 　　　**b)** $1\frac{7}{16}, 1\frac{3}{8}, \frac{3}{2}, 1.2, \frac{3}{4}$

11. Assessment Focus Amrita, Paul, and Corey baked pizzas for the fund-raising sale. The students cut their pizzas into different sized slices.

Amrita　　　　Paul　　　　Corey

Amrita sold $\frac{11}{6}$ pizzas. Paul sold 1.875 pizzas. Corey sold $\frac{9}{4}$ pizzas.

a) Use a number line to order the numbers of pizzas sold from least to greatest.

b) Who sold the most pizzas? The fewest pizzas?

c) Use a different method. Verify your answers in part b.

d) Alison sold $2\frac{1}{5}$ pizzas. Where does this fraction fit in part a?

Reflect

Describe 3 ways to compare and order fractions and decimals. Give an example of when you would use each method. Which way do you prefer? Why?

Focus	Add and subtract decimals to thousandths.

When you go to the theatre to see a movie, your attendance and how much you paid to see the movie are entered in a database.
Data are collected from theatres all across Canada and the United States.
Movie studios use these data to help predict how much money the movie will earn.

Explore

Shrek 2 was one of the highest-earning movies of 2004.
The table shows how much money *Shrek 2* earned in Canada and the United States for the first week it played in theatres.
Studios record the earnings in millions of US dollars.

Date	Earnings (US$ Millions)
Wednesday, May 19	11.786
Thursday, May 20	9.159
Friday, May 21	28.340
Saturday, May 22	44.797
Sunday, May 23	34.901
Monday, May 24	11.512
Tuesday, May 25	8.023

➤ Estimate first.
   Then find the combined earnings on:
   • the first 2 days
   • Saturday and Sunday
   • all 7 days

➤ Estimate first.
   Then find the difference in earnings on:
   • Thursday and Friday
   • Saturday and Sunday
   • Sunday and Monday
   • the days with the greatest and the least earnings

Reflect & Share

Share your results with another pair of classmates.
Discuss the strategies you used to estimate and to find the sums and differences.
Why do you think the earnings on 3 of the days are so much higher? Explain.

When we add or subtract decimals, we estimate if we do not need an exact answer.
We also estimate to check the answer is reasonable.

Example

Ephram is a long-distance runner. His practice distances
for 5 days last week are shown in the table.

a) How far did Ephram run in 5 days last week?

b) How much farther did Ephram run on Tuesday than
on Thursday?

Day	Distance (km)
Monday	8.85
Tuesday	12.25
Wednesday	10.9
Thursday	9.65
Friday	14.4

A Solution

a) $8.85 + 12.25 + 10.9 + 9.65 + 14.4$

Use front-end estimation.
Add the whole-number part of each decimal.
Think: $8 + 12 + 10 + 9 + 14 = 53$
Ephram ran about 53 km.

Add. Write each number with the same number of decimal places.
Use zeros as placeholders: 8.85, 12.25, 10.90, 9.65, 14.40
Record the numbers without the decimal points.
Add as you would whole numbers.

```
  2 3 1
    885
   1225
   1090
    965
 + 1440
   5605
```

Since the estimate is 53 km, place the decimal point after
the first 2 digits; that is, between the 6 and the 0.
Ephram ran 56.05 km.

b) Ephram ran 12.25 km on Tuesday and 9.65 km on Thursday.
Estimate.
$12.25 - 9.65$
Think: $12 - 9 = 3$
Ephram ran about 3 km farther on Tuesday.

Subtract. Align the numbers.
Subtract as you would whole numbers.

$$
\begin{array}{r}
{}^{11}\!\!\!{}^{12}\!\!\!\\
\cancel{12}.\cancel{25} \\
-\ \ 9.65 \\
\hline
2.60
\end{array}
$$

2.6 is close to the estimate 3, so the answer is reasonable.
Ephram ran 2.6 km farther on Tuesday than on Thursday.

Practice

1. Use front-end estimation to estimate each sum or difference.
   a) $2.876 - 0.975$ b) $71.382 + 6.357$
   c) $125.12 + 37.84$ d) $9.7 - 1.36$

2. The tallest building in the world is the Taipei 101 in Taiwan.
   Its height is 0.509 km. The tallest building in North America
   is the Sears Tower in Chicago, USA. Its height is 0.442 km.
   What is the difference in the heights of the buildings?

3. Four classes of students from Mackenzie School are planning
   a field trip. The total cost of the trip is $1067.50.
   To date, the classes have raised: $192.18, $212.05, $231.24, $183.77
   a) How much money have the classes raised so far?
   b) How much more money do the classes need to raise in total?
   Show your work.

4. **Assessment Focus** A baker wants to make 3 different kinds of
   chocolate chip cookies. The recipes call for 2.75 kg, 4.4 kg, and 5.55 kg
   of chocolate chips. The baker has 10.5 kg of chocolate chips.
   a) How many kilograms of chocolate chips does the baker need?
      Estimate to check your answer is reasonable.
   b) Does the baker have enough chocolate chips to make the cookies?
      How do you know?
   c) The baker wants to follow the recipes exactly.
      If your answer to part b is no, how many more kilograms of chocolate
      chips are needed? If your answer to part b is yes, how many
      kilograms of chocolate chips will the baker have left over?

5. Estimate, then calculate, the sum below.
Explain how you estimated.
$46.71 + 3.9 + 0.875$

6. The Robb family and the Chan family have similar homes.
The Robb family sets its thermostat to 20°C during the winter months.
Its monthly heating bills were: $171.23, $134.35, and $123.21
The Chan family used a programmable thermostat to lower the
temperature at night, and during the day when the family was out.
The Chan family's monthly heating bills were: $134.25, $103.27,
and $98.66

  a) How much money did each family pay to heat its home
    during the winter months?
  b) How much more money did the Robb family pay?
    Estimate to check your answer is reasonable.
  c) What other things could a family do to reduce its heating costs?

 7. Find two numbers with a difference of 151.297.

8. Use each of the digits from 0 to 7 once to make this addition true.
Find as many different answers as you can.

```
   □.□□□
 + □.□□□
 ─────────
   5 . 7 8 8
```

9. A student subtracted 0.373 from 4.81 and got the difference 0.108.
  a) What mistake did the student make?
  b) What is the correct answer?

10. Two 4-digit numbers were added. Their sum was 3.3.
What could the numbers have been?
Find as many different answers as you can. Show your work.

11. **Take It Further** Find each pattern rule. Explain how you found it.
  a) 2.09, 2.13, 2.17, 2.21, …
  b) 5.635, 5.385, 5.135, 4.885, …

How did your knowledge of estimation help you in this lesson?

Focus Use Base Ten Blocks, paper and pencil, and calculators to multiply decimals.

Recall how to multiply 2 whole numbers using Base Ten Blocks.
This picture shows the product:

$20 \times 16 = 100 + 100 + 60 + 60$
$= 320$

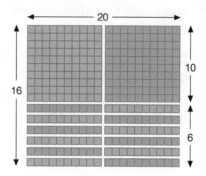

We can also use Base Ten Blocks to multiply 2 decimals.

Let the flat represent 1, the rod represent 0.1, and the small cube represent 0.01.

1   0.1   0.01

Explore

You will need Base Ten Blocks and grid paper.
Use Base Ten Blocks to model a rectangular patio with area greater than 4 m² and less than 6 m².
Let the side length of the flat represent 1 m.
How many different patios can you model?
Record your designs on grid paper.

Reflect & Share

Compare your designs with those of another pair of classmates.
Did you have any designs the same? Explain.
Explain how your designs show the area of the patio.

Connect

A rectangular park measures 1.7 km by 2.5 km.
Here are 2 ways to find the area of the park.

➤ Use Base Ten Blocks.
Build a rectangle with length 2.5 and width 1.7.
Count the blocks in the rectangle.
There are 2 flats: $2 \times 1 = 2$
There are 19 rods: $19 \times 0.1 = 1.9$
There are 35 small cubes: $35 \times 0.01 = 0.35$
The total area is: $2 + 1.9 + 0.35 = 4.25$
The total area of the park is 4.25 km².

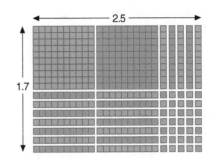

➤ Use the method for multiplying 2 whole numbers.
The area, in square kilometres, is 1.7×2.5.
Multiply: 17×25

$$\begin{array}{r} 17 \\ \times\ 25 \\ \hline 85 \\ 340 \\ \hline 425 \end{array}$$

> **1.7 × 2.5**
> Think: $1 \times 2 = 2$
> So, 1.7×2.5 is about 2.
> Place the decimal point
> between the 4 and the 2.

Using front-end estimation to place the decimal point, $1.7 \times 2.5 = 4.25$.
The area of the park is 4.25 km².

Example

At the Farmers' Market, 1 kg of grapes costs $2.95.
How much would 1.8 kg of grapes cost?

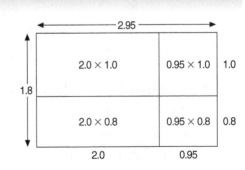

A Solution

1 kg of grapes costs $2.95.
So, 1.8 kg would cost: 2.95×1.8
Use a rectangle model.

$2.95 \times 1.8 = (2.0 \times 1.0) + (0.95 \times 1.0) + (2.0 \times 0.8) + (0.95 \times 0.8)$
$\qquad\qquad = (2 \times 1) + (0.95 \times 1) + (2 \times 0.8) + (0.95 \times 0.8)$
$\qquad\qquad = 2 + 0.95 + 1.6 + 0.76$
$\qquad\qquad = 5.31$

Another Strategy
We could use a calculator
to multiply.

1.8 kg of grapes would cost $5.31.

Use a calculator when the multiplier has more than 2 digits.

1. Write the product that each picture represents.
   Each small square represents 0.01.
   a)   b)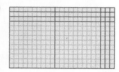

2. Use Base Ten Blocks to find each product.
   Record your work on grid paper.
   a) 2.6 × 1.5   b) 2.3 × 0.4   c) 0.8 × 0.7

3. Choose one part from question 2.
   Explain how the Base Ten Blocks show the product.

4. Multiply. Use a rectangle model.
   a) 4.2 × 3.7   b) 8.9 × 0.3   c) 0.6 × 0.9

5. A rectangular plot of land measures 30.5 m by 5.3 m.
   What is the area of the plot?
   Estimate to check your answer is reasonable.

 6. Multiply. Describe any patterns you see.
   a) 8.36 × 10   b) 8.36 × 0.1
      8.36 × 100   8.36 × 0.01
      8.36 × 1000   8.36 × 0.001
      8.36 × 10 000   8.36 × 0.0001

7. **Assessment Focus** An area rug is rectangular.
   Its dimensions are 3.4 m by 2.7 m.
   Show different strategies you can use
   to find the area of the rug.
   Which strategy is best? Justify your answer.

 8. Multiply.
   a) 2.7 × 4.786
   b) 12.52 × 13.923
   c) 0.986 × 1.352
   Explain how you can check your answers.

9. The fuel consumption estimates of Josie's car are:

City: 21.2 km/L   Highway: 23.3 km/L

The car's gas tank holds 40.2 L of fuel.

a) How far could Josie drive on a full tank of gas on the highway before she runs out of fuel?

b) How far could she drive on a full tank of gas in the city? What assumptions did you make?

10. Find the cost of each item at the Farmers' Market. Which strategy will you use? Justify your choice.

a) 2.56 kg of apples at $0.95/kg

b) 10.5 kg of potatoes at $1.19/kg

c) 0.25 kg of herbs at $2.48/kg

11. The product of 2 decimals is 0.36. What might the decimals be? Find as many answers as you can.

12. a) Multiply 18 × 12.

b) Use only the result from part a and estimation. Find each product.

i) 1.8 × 12   ii) 18 × 0.12   iii) 0.18 × 12   iv) 0.18 × 0.12

Explain your strategies.

13. Take It Further

a) Multiply.

i) 6.3 × 1.8   ii) 0.37 × 0.26   iii) 3.52 × 2.4   iv) 1.234 × 0.9

b) Look at the questions and products in part a.

What patterns do you see in the numbers of decimal places in the question and the product?

How could you use this pattern to place the decimal point in a product without estimating?

c) Multiply: 2.6 × 3.5

Does the pattern from part b hold true?

If your answer is no, explain why not.

Reflect

When you multiply 2 decimals, how do you know where to place the decimal point in the product? Use examples to explain.

3.5 Dividing Decimals

Focus Use Base Ten Blocks, paper and pencil, and calculators to divide decimals.

Recall how you used Base Ten Blocks to multiply:

Since multiplication and division are related, we can also use Base Ten Blocks to divide.

Which division sentences could you write for this diagram?

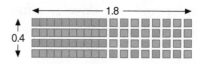

$$1.8 \times 0.4 = 0.72$$

Explore

You will need Base Ten Blocks and grid paper.
Marius bought 1.44 m of ribbon for his craft project.
He needs to cut the ribbon into 0.6-m lengths.
How many 0.6-m lengths can he cut?
Use Base Ten Blocks to find out.
Record your work on grid paper.

Reflect & Share

Compare your solution with that of another pair of classmates.
What was your strategy?
How could you use division of whole numbers to check your answer?

Connect

Jan bought 2.8 m of framing to make picture frames.
Each picture needs 0.8 m of frame.
How many frames can Jan make?
How much framing material is left over?

Use Base Ten Blocks to divide: $2.8 \div 0.8$

Make a rectangle with area 2.8 and width 0.8.

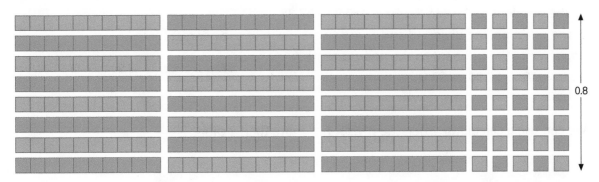

The length of the rectangle is 3.5.

So, Jan can make 3 frames.

3 frames use: 3×0.8 m $= 2.4$ m

So, the framing material left is: 2.8 m $-$ 2.4 m $= 0.4$ m

Sometimes when we divide 2 decimals, the quotient is not a terminating decimal.
Then we can use paper and pencil.

Example

Divide: $52.1 \div 0.9$

A Solution

Estimate first: $52.1 \div 0.9$

Write each decimal to the nearest whole number, then divide.

$52 \div 1 = 52$

So, $52.1 \div 0.9$ is about 52.

> 52.1 is closer to 52.
> 0.9 is closer to 1.

Divide as you would whole numbers.

$521 \div 9$

```
           5788  ←—— quotient
divisor —→ 9)52100  ←—— dividend
           45
           ‾‾
           71
           63
           ‾‾
           80
           72
           ‾‾
           80
           72
           ‾‾
            8
```

> Divide until the quotient has 2 more digits than the estimate. Then we can write the quotient to the nearest tenth.

If the quotient is not exact, write zeros in the dividend, then continue to divide.

Since the estimate has 2 digits, divide until there are 4 digits in the quotient.

Since the estimate was 52, place the decimal point
so the quotient is close to 52: $52.1 \div 0.9 \doteq 57.88$
In the question, the dividend and divisor were given to the nearest tenth.
So, we write the quotient to the nearest tenth.
$52.1 \div 0.9 \doteq 57.90$, or 57.9

57.88 is closer to 57.90 than to 57.80.

We can use a calculator when the divisor has more than 1 digit.

Practice

1. Use Base Ten Blocks to divide. Record your work on grid paper.
   a) $0.8 \div 0.1$ b) $1.2 \div 0.3$ c) $2.7 \div 0.6$ d) $2.2 \div 0.4$

 2. Divide. Describe any patterns you see.
   a) $124.5 \div 10$ b) $124.5 \div 0.1$
      $124.5 \div 100$ $124.5 \div 0.01$
      $124.5 \div 1000$ $124.5 \div 0.001$
      $124.5 \div 10\ 000$ $124.5 \div 0.0001$

3. Why do all these division statements have 6 as the answer?
   a) $30 \div 5$ b) $3.0 \div 0.5$ c) $0.3 \div 0.05$ d) $300 \div 50$
   Which one is easiest to calculate? Explain.

4. Estimate to choose the correct quotient for each division question.

Question	Possible Quotients		
a) $59.5 \div 5$	119	11.9	1.19
b) $195.3 \div 0.2$	9765	976.5	97.65
c) $31.32 \div 0.8$	3915	391.5	39.15

5. Use paper and pencil to divide.
   a) $1.5 \div 0.6$ b) $2.24 \div 0.7$ c) $1.28 \div 0.8$ d) $2.16 \div 0.9$

 6. Divide. Write each quotient to the nearest tenth.
   Use front-end estimation to check your answer is reasonable.
   a) $8.36 \div 2.4$ b) $1.98 \div 1.3$ c) $27.82 \div 3.9$ d) $130.4 \div 5.4$

7. A toonie is approximately 0.2 cm thick.
   How many toonies are in a stack of toonies 17.4 cm high?

8. The area of a large rectangular flowerbed is 22.32 m². The width is 0.8 m. What is the length?

9. A 0.4-kg bag of oranges costs $1.34.
   a) Estimate. About how much does 1 kg of oranges cost?
   b) What is the actual cost of 1 kg of oranges? How do you know your answer is reasonable?
   c) Suppose you spent $10 on oranges. What mass of oranges did you buy?

10. **Assessment Focus** Alex finds a remnant of landscaping fabric at a garden store. The fabric is the standard width, with length 9.88 m. Alex needs fourteen 0.8-m pieces for a garden patio.
   a) How many 0.8-m pieces can Alex cut from the remnant? What assumptions did you make?
   b) Will Alex have all the fabric he needs? Why or why not?
   c) If your answer to part b is no, how much more fabric does Alex need?
   d) Alex redesigns his patio so that he needs fourteen 0.7-m pieces of fabric. Will the remnant be enough fabric? Explain.

11. The quotient of two decimals is 0.12. What might the decimals be? Write as many different possible decimal pairs as you can.

12. Last week, Alicia worked 37.5 h. She earned $346.88. How much money did Alicia earn per hour? Why is the answer different from the number in the calculator display?

13. The question 237 ÷ 7 does not have an exact quotient. The first five digits of the quotient are 33857. The decimal point has been omitted. Use only this information and estimation. Write an approximate quotient for each question. Justify each answer.
   a) 237 ÷ 0.7   **b)** 2.37 ÷ 0.07   **c)** 23.7 ÷ 7   **d)** 2370 ÷ 70

Reflect

Talk to a partner. Tell how you can find 1.372 ÷ 0.7 by dividing by 7. Why does this work?

3.6 Order of Operations with Decimals

Explore

How many different ways can you find the answer for this expression?

$6 \times 15.9 + 36.4 \div 4$

Show your work for each answer.

Reflect & Share

Compare your answers with those of another pair of classmates.
Which solution do you think is correct? Explain your reasoning.

Connect

To make sure everyone gets the same answer for a given expression,
we add, subtract, multiply, and divide in this order:
- Do the operations in brackets first.
- Then divide and multiply, in order, from left to right.
- Then add and subtract, in order, from left to right.
When we find the answer to an expression, we *evaluate*.

> We use the same order of operations for decimals as for whole numbers.

Example

Evaluate: $12.376 \div (4.75 + 1.2) + 2.45 \times 0.2 - 1.84$

> Use a calculator when you need to.

A Solution

$12.376 \div (4.75 + 1.2) + 2.45 \times 0.2 - 1.84$ Calculate in brackets.

$= 12.376 \div 5.95 + 2.45 \times 0.2 - 1.84$ Multiply and divide from left to right.

$= 2.08 + 0.49 - 1.84$ Add and subtract from left to right.

$= 2.57 - 1.84$

$= 0.73$

Many calculators follow the order of operations.

To see whether your calculator does, enter: $12.4 \times 2.2 - 15.2 \div 4$

If your answer is 23.48, your calculator follows the order of operations.

Practice

1. Evaluate.

    a) $4.6 + 5.1 - 3.2$ **b)** $8 - 3.6 \div 2$ **c)** $46.4 - 10.8 \times 3$ **d)** $85.6 \div 0.4 \times 7$

2. Evaluate.

    a) $(46.78 - 23.58) \times 2.5$ **b)** $(98.5 + 7) \div 0.5$ **c)** $7.2 \div (2.4 - 1.8)$

3. Evaluate.

    a) $9.8 - 3.2 \div 0.4 + 2.6$ **b)** $(9.8 - 3.2) \div (0.4 + 2.6)$

    Explain why the answers are different.

4. Evaluate.

    a) $1.35 + (5 \times 4.9 \div 0.07) - 2.7 \times 2.1$ **b)** $9.035 \times 5.2 - 4.32 \times 6.7$

    c) $2.368 \div 0.016 + 16.575 \div 1.105$ **d)** $0.38 + 16.2 \times (2.1 + 4.7) + 21 \div 3.5$

5. **Assessment Focus** Ioana, Aida, and Norman got different answers

    for this problem: $12 \times (4.8 \div 0.3) - 3.64 \times 3.5$

    Ioana's answer was 39.12, Aida's answer was 179.26,

    and Norman's answer was 659.26.

    a) Which student had the correct answer?

        How do you know?

    b) Show and explain how the other

        two students got their answers.

        Where did they go wrong?

6. Evaluate. Show all steps:

    $0.38 + 16.2 \times (2.1 - 1.2) + 21 \div 0.8$

7. **Take It Further** Use at least 4 of the numbers 0.1, 0.2, 0.3, 0.4, 0.5, 0.6, 0.7, 0.8 and 0.9,

    and any operations or brackets to make each whole number from 1 to 5.

Reflect

    Why do we need to agree on an order of operations?

Mid-Unit Review

LESSON

Use a calculator when you need to.

3.1 **1. a)** Write each fraction as a decimal.

i) $\frac{1}{33}$ ii) $\frac{2}{33}$ iii) $\frac{3}{33}$

b) Describe the pattern in your answers to part a.

c) Use your pattern to predict the fraction form of each decimal.

i) $0.\overline{15}$ ii) $0.\overline{36}$

2. Write each fraction as a decimal. Identify the decimals as repeating or terminating.

a) $\frac{1}{8}$ **b)** $\frac{3}{5}$

c) $\frac{2}{3}$ **d)** $\frac{7}{13}$

3. Write each decimal as a fraction.

a) 0.2   **b)** $0.\overline{8}$

c) 0.005   **d)** $0.\overline{23}$

3.2 **4.** Order each set of numbers from least to greatest. Use a different method for each part.

a) $2\frac{1}{4}, \frac{11}{6}, \frac{8}{3}, 2$ **b)** $3.5, \frac{23}{8}, 1\frac{3}{4}$

c) $1.75, \frac{13}{10}, \frac{9}{5}, 1\frac{3}{5}, 1$

5. Find a number between each pair of numbers. Which strategy did you use each time?

a) $\frac{4}{3}, \frac{5}{3}$ **b)** $2\frac{3}{8}, \frac{5}{2}$ **c)** $1.4, \frac{8}{5}$

3.3 **6.** Use front-end estimation to place the decimal point in each answer.

a) $32.47 - 6.75 = 2\,5\,7\,2$

b) $118.234 + 19.287 = 1\,3\,7\,5\,2\,1$

c) $17.9 - 0.8 = 1\,7\,1$

7. Winsome is being trained as a guide dog for a blind person. At birth, she had a mass of 0.475 kg. At 6 weeks, her mass was 4.06 kg. From 6 weeks to 12 weeks, she gained 5.19 kg.

a) By how much did Winsome's mass change from birth to 6 weeks?

b) What was her mass at 12 weeks?

3.4 **8.** Estimate to place the decimal point in each product. Show your estimation strategy.

a) $9.3 \times 0.8 = 7\,4\,4$

b) $3.62 \times 1.3 = 4\,7\,0\,6$

c) $11.25 \times 5.24 = 5\,8\,9\,5$

9. A rectangular park has dimensions 2.84 km by 3.5 km. What is the area of the park?

3.5 **10.** When you divide 15.4 by 2, the quotient is 7.7. When you divide 1.54 by 0.2, the quotient is 7.7. Explain why the quotients are the same.

3.6 **11.** Evaluate.

a) $5.9 + 3.7 \times 2.8$

b) $12.625 \times (1.873 + 2.127)$

c) $2.1 \div 0.75 + 6.38 \times 2.45$

Relating Fractions, Decimals, and Percents

Focus | Relate percent to fractions and decimals.

We see uses of percent everywhere.

What do you know from looking at each picture?
Recall that percent means per hundred.
49% is $\frac{49}{100}$ = 0.49

Explore

Your teacher will give you a large copy of this puzzle.
Describe each puzzle piece as a percent, then as a fraction
and a decimal of the whole puzzle.

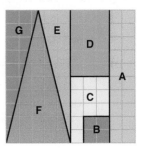

Reflect & Share

Compare your answers with those of another pair of classmates.
If the answers are different, how do you know which are correct?

Connect

➤ We can use number lines
to show how percents
relate to fractions
and decimals.
For example:
25% = $\frac{25}{100}$ = 0.25

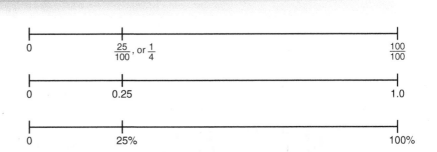

➤ Conversely, a decimal
can be written as a percent:
0.15 = $\frac{15}{100}$ = 15%

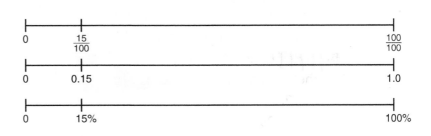

➤ To write a fraction as a percent, write the equivalent fraction with denominator 100.
For example:

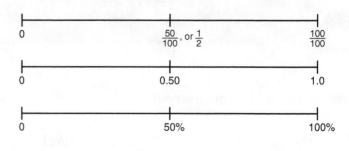

$$\overset{\times\ 50}{\frac{1}{2}} = \frac{50}{100} = 50\%$$
$$\underset{\times\ 50}{}$$

Example

a) Write each percent as a fraction and as a decimal.

   i) 75%   ii) 9%

b) Write each fraction as a percent and as a decimal.

   i) $\frac{2}{5}$ ii) $\frac{7}{20}$

Draw number lines to show how the numbers are related.

A Solution

a) i) $75\% = \frac{75}{100} = 0.75$

    ii) $9\% = \frac{9}{100} = 0.09$

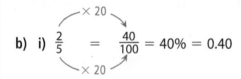

b) i) $\overset{\times\ 20}{\frac{2}{5}} = \frac{40}{100} = 40\% = 0.40$

    ii) $\overset{\times\ 5}{\frac{7}{20}} = \frac{35}{100} = 35\% = 0.35$

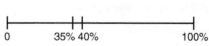

Practice

1. What percent of each hundred chart is shaded?
Write each percent as a fraction and as a decimal.

a)    b) 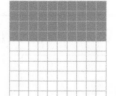   c)

2. Write each percent as a fraction and a decimal.
Sketch number lines to show how the numbers are related.
a) 2%　　　　　**b)** 9%　　　　　**c)** 28%　　　　　**d)** 95%

3. Write each fraction as a decimal and a percent.
a) $\frac{2}{10}$　　　　**b)** $\frac{3}{50}$　　　　**c)** $\frac{4}{25}$　　　　**d)** $\frac{13}{20}$　　　　**e)** $\frac{4}{5}$.

4. Fred had 8 out of 10 on a test. Janet had 82% on the test.
Who did better? How do you know?

5. **Assessment Focus** You will need a sheet of paper
and coloured pencils.
Divide the paper into these 4 sections.
- 1 blue section that is $\frac{1}{2}$ of the page
- 1 red section that is 10% of the page
- 1 yellow section that is 25% of the page
- 1 green section to fill the remaining space.

Explain how you did this.
What percent of the page is the green section?
How do you know?

6. **Take It Further** Suppose each pattern is continued on a hundred chart.
The numbers in each pattern are coloured red.
For each pattern, what percent of the numbers on the chart are red?
Explain your strategy for each pattern.
a) 4, 8, 12, 16, 20, …　　**b)** 1, 3, 5, 7, …　　　**c)** 2, 4, 8, 16, …　　　**d)** 1, 3, 7, 13, …

Reflect

Suppose you know your mark out of 20 on an English test.
Tell how you could write the mark as a decimal and a percent.

3.8 Solving Percent Problems

Focus Solve problems involving percents to 100%.

When shopping, it is often useful to be able to calculate a percent, to find the sale price, the final price, or to decide which of two offers is the better deal.

Explore

A jacket originally cost $48.00.
It is on sale for 25% off.
What is the sale price of the jacket?
How much is saved by buying
the jacket on sale?
Find several ways to solve this problem.

Reflect & Share

Compare strategies with those of another pair of classmates.
Which strategy would you use if the sale was 45% off? Explain your choice.

Connect

A paperback novel originally cost $7.99.
It is on sale for 15% off.
To find how much you save, calculate 15% of $7.99.

$15\% = \frac{15}{100} = 0.15$

So, 15% of $7.99 = $\frac{15}{100}$ of 7.99

$= 0.15 \times 7.99$

Use a calculator.

$0.15 \times 7.99 = 1.1985$

So, $0.15 \times \$7.99 = \$1.1985$

$1.1985 to the nearest cent is $1.20.
You save $1.20 by buying the book on sale.
We can show this on a number line.

Estimate to check if the answer is reasonable.

15% is about 20%, which is $\frac{1}{5}$.

$7.99 is about $10.00.

So, 0.15×7.99 is about $\frac{1}{5}$ of 10, which is 2.

This is close to the calculated amount, so the answer is reasonable.

Example

Sandi works at Fancies Flowers on Saturdays.
The owner pays Sandi 3% of all money she takes in on a day.
Last Saturday, Sandi took in $1200.00.
How much money did Sandi earn last Saturday?
Illustrate the answer on a number line.

A Solution

Sandi took in $1200.00.

We want to find 3% of $1200.00.

3% is $\frac{3}{100} = 0.03$

So, 3% of $1200 = 0.03 \times 1200$

Ignore the decimal point and multiply as whole numbers.

Another Strategy
We could find 1% of $1200.00,
then multiply by 3.

Estimate to place the decimal point.
$1200 is about $1000.
1% of $1000 is $10.
So, 3% of $1000 is: $10 \times 3 = $30

$$\begin{array}{r} 1200 \\ \times\ \ 3 \\ \hline 3600 \end{array}$$

So, $0.03 \times $1200 = 36.00

Sandi earned $36.00 last Saturday.

Show this on a number line.

Practice

1. Calculate.

   a) 10% of 30   **b)** 20% of 50   **c)** 18% of 36   **d)** 67% of 112

2. The regular price of a radio is $60.00.
   Find the sale price before taxes when the radio is on sale for:

   a) 25% off   **b)** 30% off   **c)** 40% off

3. Find the sale price before taxes of each item.

a) coat: 55% off $90   b) shoes: 45% off $40   c) sweater: 30% off $50

4. Find the tip left by each customer at a restaurant.

a) Denis: 15% of $24.20   b) Molly: 20% of $56.50   c) Tudor: 10% of $32.70

5. The Goods and Services tax (GST) is currently 6%.
For each item below:

i) Find the GST.

ii) Find the cost of the item including GST.

a) bicycle: $129.00   b) DVD: $24.99   c) skateboard: $42.97

6. There are 641 First Nations bands in Canada.
About 30% of these bands are in British Columbia.
About how many bands are in British Columbia?
Sketch a number line to show your answer.

7. **Assessment Focus** A clothing store
runs this advertisement in a local paper.
"Our entire stock up to 60% off"

a) What does "up to 60% off" mean?

b) Which items in the advertisement
have been reduced by 60%?

c) Suppose all items are reduced by 60%.
Explain the changes you would make
to the sale prices.

Item	Regular Price	Sale Price
Sweaters	$49.99	$34.99
Ski Jackets	$149.99	$112.49
Scarves	$29.99	$12.00
Leather Gloves	$69.99	$38.49
Hats	$24.99	$10.00

8. **Take It Further** Marissa and Jarod plan to purchase
DVD players with a regular price of $199.99.
The DVD players are on sale for 25% off.
Marissa starts by calculating 25% of $199.99.
Jarod calculates 75% of $199.99.

a) Show how Marissa uses her calculation to find the sale price.

b) How does Jarod find the sale price? Show his work.

c) Do both methods result in the same sale price? Explain.

Reflect

How does a good understanding of percents help you outside
the classroom? Give an example.

Sports Trainer

Sports trainers use scientific research and scientific techniques to maximize an athlete's performance. An athlete may be measured for percent body fat, or percent of either fast- or slow-twitch muscle fibre.

A trainer may recommend the athlete eat pre-event meals that contain a certain percent of carbohydrate, or choose a "sports drink" that contains a high percent of certain minerals. The trainer creates and monitors exercise routines. These enable the athlete to attain a certain percent of maximum heart rate, speed, or power.

Most sports drinks contain minerals. Research shows that the most effective sports drink has a magnesium to calcium ratio of 1:2. The body absorbs about 87% of magnesium in a drink, and about 44% of calcium in a drink. One serving of a particular sports drink contains about 96 mg of calcium and 48 mg of magnesium. About how many milligrams of each mineral will the body absorb?

Writing Instructions

Life is full of instructions.
If you have ever filled out a form,
assembled a desk, or followed
directions to someone's house,
you know the importance of
good instructions.

Rhonda describes
this shape to Rashad.
She asks Rashad
to sketch it.

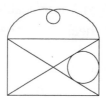

Ok, draw a box, then make an "x" in the centre of the box. Draw a small circle on the right. There is a curly line on top, so draw a line from one corner to the other that loops once in the middle.

- Follow Rhonda's instructions.
  Draw 3 different shapes that are different from the original figure.
- Describe how to improve Rhonda's instructions.
- Rewrite Rhonda's instructions using your suggestions.

Instructions for Drawing

- Draw a shape.
  Write instructions that someone else could follow to draw it.
- Trade instructions with a classmate.
  Follow your classmate's instructions to draw the shape.
- Compare your shape with the original shape.
  Are the shapes the same?
  If not, suggest ways to improve your classmate's instructions.

Instructions for Calculations

- Your friend has forgotten his calculator at school.
  Write instructions that your friend could follow
  to find 87×0.17 without a calculator.
- Trade instructions with a classmate.
  Follow your classmate's instructions
  to find the product.
- Use a calculator to find the product.
  Are the products the same?
  If not, suggest ways to improve
  your classmate's instructions.

Unit Review

What Do I Need to Know?

✓ Here are some fractions and decimals you should know.

$\frac{1}{2} = 0.5$ $\qquad$ $\frac{1}{3} = 0.\overline{3}$ $\qquad$ $\frac{1}{4} = 0.25$

$\frac{1}{5} = 0.2$ $\qquad$ $\frac{1}{8} = 0.125$ $\qquad$ $\frac{1}{10} = 0.1$

$\frac{1}{20} = 0.05$ $\qquad$ $\frac{1}{100} = 0.01$ $\qquad$ $\frac{1}{1000} = 0.001$

✓ Percent is the number of parts per hundred.
A percent can be written as a fraction and as a decimal.

✓ Here are some fractions, decimals, and percents that you should also know.

$\frac{1}{4} = 0.25 = 25\%$ $\qquad$ $\frac{1}{2} = 0.5 = 50\%$ $\qquad$ $\frac{3}{4} = 0.75 = 75\%$

$\frac{1}{20} = 0.05 = 5\%$ $\qquad$ $\frac{1}{10} = 0.1 = 10\%$ $\qquad$ $\frac{1}{5} = 0.2 = 20\%$

✓ The order of operations with whole numbers applies to decimals.
  • Do the operations in brackets first.
  • Then divide and multiply, in order, from left to right.
  • Then add and subtract, in order, from left to right.

Your World
When you buy a Canada Savings
Bond (CSB), you are lending
the Canadian government money.
The government pays you
for borrowing your money.
It pays a percent of what you invested.
In 2006, if you invested money for 1 year,
the government paid you 2%
of the amount you invested.
Suppose you bought a $2000 CSB in 2006.
How much will the government pay you for 1 year?
What if you bought a $2500 CSB?

What Should I Be Able to Do?

LESSON

3.1

1. Write each fraction as a decimal. Identify each decimal as terminating or repeating.

a) $\frac{3}{5}$ b) $\frac{5}{6}$ c) $\frac{3}{8}$ d) $\frac{3}{20}$

2. Write each decimal as a fraction or a mixed number in simplest form.

a) 0.55   b) $1.\overline{3}$
c) 0.8   d) $0.\overline{07}$

3.2

3. a) Use any method. Order these numbers from least to greatest. Explain the method you used.

$\frac{5}{4}, 1\frac{1}{16}, \frac{3}{6}, 1.1, \frac{5}{8}$

b) Use a different method to order the numbers, to verify your answer in part a.

4. In each ordered set, identify the number that has been placed incorrectly. Explain how you know.

a) $2\frac{1}{3}, 2.25, \frac{17}{6}, 2\frac{11}{12}$

b) $\frac{3}{5}, \frac{9}{10}, \frac{21}{20}, 1\frac{3}{15}, 1.1$

3.3

5. Two decimals have a sum of 3.41. What might the decimals be? Find as many answers as you can.

6. Asafa Powell of Jamaica holds the men's world record for the 100-m sprint, with a time of 9.77 s. Florence Griffith Joyner of the United States holds the women's world record, with a time of 10.49 s. What is the difference in their times?

3.4 **7.** Kiah works at the library after school. She earns $7.65/h. She usually works 15.5 h a week.

a) What does Kiah earn in a week? Use estimation to check your answer.

b) One week Kiah only works one-half the hours she usually works. What are her earnings that week?

8. Lok needs 1.2 m of fabric to make a tote bag. He finds two fabrics he likes. One fabric costs $7.59/m and the other fabric costs $6.29/m. How much money will Lok save if he buys the less expensive fabric?

3.5 **9.** Estimate.
Which quotients are:
 i) greater than 100?
ii) less than 50?
Calculate the quotients that are less than 50.

a) $259.8 \div 1.65$

b) $35.2 \div 0.2$

c) $175.08 \div 0.8$

d) $93.8 \div 22.4$

e) $162.24 \div 31.2$

f) $883.3 \div 36.5$

10. The area of a rectangle is 3.75 m². Its length is 0.6 m. What is the width of the rectangle?

 11. Evaluate. Use the order of operations.
a) $8.11 + 6.75 \times 5.6 - 2.12$
b) $3.78 \times 2.25 - 4.028 \div 1.52$

12. a) Simplify.
i) $1.2 + 2.8 \times 2.1 + 3.6$
ii) $1.2 \times 2.8 + 2.1 \times 3.6$
iii) $1.2 \times (2.8 + 2.1) + 3.6$
iv) $1.2 + 2.8 + 2.1 \times 3.6$
b) All the expressions in part a have the same numbers and operations. Why are the answers different?

3.7 13. Write each percent as a fraction and as a decimal. Sketch number lines to illustrate.
a) 80%   b) 12%
c) 2%   d) 63%

14. Write each fraction as a decimal and as a percent. Sketch number lines to illustrate.
a) $\frac{14}{25}$ b) $\frac{19}{20}$
c) $\frac{7}{50}$ d) $\frac{1}{5}$

3.8 15. There are 35 students in a Grade 7 class. On one day, 20% of the students were at a sports meet. How many students were in class?

16. Find the sale price before taxes of each item.
a) video game: 15% off $39
b) lacrosse stick: 25% off $29
c) fishing rod: 30% off $45

17. A souvenir Olympic hat sells for $29.99.
a) Russell lives in Newfoundland where there is a sales tax of 14%. Calculate the final cost of the hat in Newfoundland.
b) Jenna lives in Alberta where the GST tax is 6%. Calculate the final cost of the hat in Alberta.
c) What is the difference between the final costs of the hat in Newfoundland and Alberta?

18. Madeleine received good service in a restaurant. She left the waitress a tip of 20%. Madeleine's bill was $32.75. How much tip did the waitress receive? Show your work. Draw a number line to illustrate your answer.

Practice Test

1. Write each decimal as a fraction in simplest form and each fraction as a decimal.
   a) 0.004
   b) 0.64
   c) $0.\overline{3}$
   d) $\frac{51}{200}$
   e) $\frac{3}{4}$

2. Ryan earns $18.00 a day walking dogs.
   He walked dogs 5 days last week.
   a) How much money did Ryan earn last week?
   Ryan is saving to buy inline skating equipment.
   The skates cost $59.95.
   A helmet costs $22.90.
   A set of elbow, knee, and wrist guards costs $24.95.
   b) Does Ryan have enough money to buy the equipment?
   Show your work.
   c) If your answer to part b is no, how much more money does Ryan need?
   What assumptions did you make?

3. Maria stated that $1\frac{5}{6}$ is between 1.8 and $\frac{13}{7}$.
   Do you agree?
   Give reasons for your answer.

4. Evaluate.
   a) $3.8 + 5.1 \times 6.4 - 1.7$
   b) $3.54 \div 0.3 + (2.58 \times 1.5)$

5. Last spring, 40 cats were adopted from the local animal shelter.
   This spring, the number of cats adopted dropped by 35%.
   How many cats were adopted this spring?
   Draw a number line to show your answer.

6. The regular price of a pair of shoes is $78.00.
   The shoes are on sale for 25% off.
   a) What is the sale price of the shoes?
   b) How much money is saved by buying the shoes on sale?
   c) The GST is 6%. How much GST would be added to the sale price?
   d) What is the final price of the shoes?

Part 1

Tanya and Marcus used the store's coupon below while shopping at *Savings for U*. They purchased:

a) Find the total cost of the items before sales tax.

b) Find the total cost of the items after they used the coupon.

c) Find the total cost of the items including GST of 6%.

Part 2

Winnie used the Scratch'n'Save coupon to buy:

The clerk scratched the coupon to reveal 20%.

a) Find the total cost of the items before sales tax.

b) Find the total cost of the items after the scratch coupon was used.

c) Find the total cost of the items including GST of 6%.

Part 3

Marty wants to purchase the following items.

i) 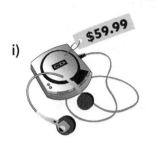 $59.99

ii)  $85.99

iii)  $19.95

iv)  $37.95

Use each discount coupon on page 124. The scratch coupon shows 20%.
Calculate the cost of each item, with each coupon.
Which coupon offers the better deal on each item?

Use each discount coupon on page 124.

Check List

Your work should show:

✓ all calculations in detail

✓ that you can use percents to solve real-life problems

✓ clear explanations and conclusions, and how you reached your conclusions

Reflect on Your Learning

Look back at the goals under *What You'll Learn*.
Which goals were easiest for you to achieve? Why do you think so?
Which were more challenging? What strategies did you use to meet these goals?

UNIT

1

1. Copy this Venn diagram.
Sort these numbers.

| 320 | 264 | 762 | 4926 |
| 2660 | 1293 | 488 | 504 |

How did you know where to put
each number?

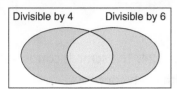

2. Suppose you have 40 strawberries.
You must share the strawberries
equally with everyone at the picnic
table. How many strawberries will
each person get, in each case?
a) There are 8 people at the table.
b) There are 5 people at the table.
c) There is no one at the table.
Explain your answer.

3. Write an algebraic expression for
each phrase.
a) a number divided by twelve
b) eleven added to a number
c) eight less than a number

4. a) Describe the patterns in this table.
b) Use the patterns to extend the
table 3 more rows.
c) Use algebra.
Write a relation
that describes
how the output
is related to
the input.

Input x	Output
1	4
2	6
3	8
4	10

5. Identify the numerical coefficient,
the variable, and the constant term
in each algebraic expression.
a) $3s + 2$ **b)** $7p$
c) $c + 8$ **d)** $11w + 9$

6. The cost to park a car is $5 for
the first hour, plus $3 for each
additional half hour.
a) Write a relation to show how
the total cost is related to the
number of additional half hours.
b) Copy and complete this table.

Number of Additional Half Hours	Cost ($)
0	
1	
2	
3	
4	

c) Draw a graph to show the
relation. Describe the graph.
d) Use the graph to answer
these questions.
  i) Tanya parked for 6 additional
  half hours.
  What was her total cost?
  ii) Uton paid $29 to park his car.
  How long was he parked?

7. Draw pictures to represent the steps
you took to solve each equation.
a) $3x = 15$
b) $x + 9 = 11$

8. a) Suppose you have 8 yellow tiles, and use all of them. How many red tiles would you need to model −3? How do you know?

b) Suppose you have 5 red tiles and 5 yellow tiles. How many ways can you find to model −3 with tiles?

9. Use coloured tiles to represent each sum. Find each sum. Sketch the tiles you used.

a) $(-7) + (+7)$ **b)** $(-7) + (+5)$

c) $(-7) + (-5)$ **d)** $(+7) + (-5)$

10. Use a number line. For each sentence below:

a) Write each number as an integer.

b) Write an addition equation. Explain your answer in words.

  i) You deposit $10, then withdraw $5.

  ii) A balloon rises 25 m, then falls 10 m.

  iii) You ride the elevator down 9 floors, then up 12 floors.

11. What is the difference in altitudes? How can you subtract to find out?

a) An altitude of 80 m above sea level and an altitude of 35 m below sea level

b) An altitude of 65 m below sea level and an altitude of 10 m above sea level

12. Add or subtract.

a) $(+5) + (-9)$ **b)** $(-1) + (-5)$

c) $(+2) - (-8)$ **d)** $(-9) - (-3)$

13. a) Write each fraction as a decimal.

  i) $\frac{1}{33}$ **ii)** $\frac{2}{33}$ **iii)** $\frac{3}{33}$

b) Describe the pattern in your answers to part a.

c) Use your pattern to predict the fraction form of each decimal.

  i) $0.\overline{15}$ **ii)** $0.\overline{24}$ **iii)** $0.\overline{30}$

14. a) Use any method. Order these numbers from greatest to least.

$\frac{21}{4}, 4.9, 5\frac{1}{3}, \frac{24}{5}, 5.3$

b) Use a different method. Verify your answer in part a.

15. The tallest woman on record was 2.483 m tall. The shortest woman on record was 0.61 m tall. What is the difference in their heights?

16. Multiply. Draw a diagram to show each product.

a) 2.3×3.4 **b)** 1.8×2.2

c) 4.1×3.7 **d)** 1.7×2.9

17. Nuri has 10.875 L of water. He pours 0.5 L into each of several plastic bottles.

a) How many bottles can Nuri fill?

b) How much water is left over?

18. The Goods and Services Tax (GST) is currently 6%. For each item below:

  i) Find the GST.

  ii) Find the cost of the item including GST.

    a) snowshoes that cost $129.99

    b) a CD that costs $17.98

Circles and Area

Look at these pictures.
What shapes do you see?

- What is a circle?
- Where do you see circles?
- What do you know about a circle?
- What might be useful to know
  about a circle?
  A parallelogram?
  A triangle?

What You'll Learn

- Investigate and explain the relationships among the radius, diameter, and circumference of a circle.
- Determine the sum of the central angles of a circle.
- Construct circles and solve problems involving circles.
- Develop formulas to find the areas of a parallelogram, a triangle, and a circle.
- Draw, label, and interpret circle graphs.

Why It's Important

- The ability to measure circles, triangles, and parallelograms is an important skill.
  These shapes are used in design, architecture, and construction.

Key Words

- radius, radii
- diameter
- circumference
- π
- irrational number
- base
- height
- circle graph
- sector
- legend
- percent circle
- central angle
- sector angle
- pie chart

Investigating Circles

Explore

You will need circular objects, a compass, and a ruler.

➤ Use a compass. Draw a large circle. Use a ruler.
Draw a line segment that joins two points
on the circle. Measure the line segment.
Label the line segment with its length.
Draw and measure other segments
that join two points on the circle.
Find the longest segment in the circle.
How many other segments can you draw with
this length? Repeat the activity for other circles.

➤ Trace a circular object. Cut out the circle.
How many ways can you find the centre of the circle?
Measure the distance from the centre to the circle.
Measure the distance across the circle, through its centre.
Record the measurements in a table.
Repeat the activity with other circular objects.
What pattern do you see in your results?

Reflect & Share

Compare your results with those of another pair of classmates.
Where is the longest segment in any circle?
What relationship did you find between the distance across a circle
through its centre, and the distance from the centre to the circle?

Connect

All points on a circle are the same
distance from the centre of the circle.
This distance is the **radius** of the circle.

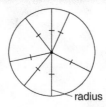

radius

The longest line segment in any circle is the **diameter** of the circle. The diameter passes through the centre of the circle. The radius is one-half the length of the diameter. The diameter is two times the length of the radius.

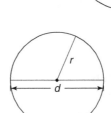

Let r represent the radius, and d the diameter. Then the relationship between the radius and diameter of a circle is:

$r = d \div 2$, which can be written as $r = \frac{d}{2}$

And, $d = 2r$

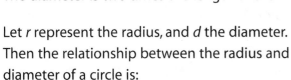

> The plural of *radius* is *radii*; that is, one radius, two or more radii.

Example

Use a compass. Construct a circle with:

a) radius 5 cm   **b)** diameter 10 cm

What do you notice about the circles you constructed?

A Solution

a) Draw a line segment with length 5 cm.
Place the compass point at one end.
Place the pencil point at the other end.
Draw a circle.

b) Draw a line segment with length 10 cm.
Use a ruler to find its midpoint.
Place the compass point at the midpoint.
Place the pencil point at one end of the segment.
Draw a circle.

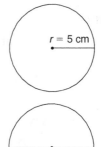

The two circles are congruent.
A circle with radius 5 cm has diameter 10 cm.

> Recall that congruent shapes are identical.

Practice

1. Use a compass.
Draw a circle with each radius.
a) 6 cm   **b)** 8 cm
Label the radius, then find the diameter.

2. Draw a circle with each radius without using a compass.

   a) 7 cm   **b)** 4 cm

   Label the radius, then find the diameter.

   Explain the method you used to draw the circles.

   What are the disadvantages of not using a compass?

3. a) A circle has diameter 3.8 cm. What is the radius?

   b) A circle has radius 7.5 cm. What is the diameter?

4. A circular tabletop is to be cut from a rectangular piece
of wood that measures 1.20 m by 1.80 m.
What is the radius of the largest tabletop that could be cut?
Justify your answer. Include a sketch.

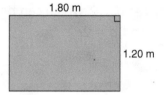

1.80 m

1.20 m

5. a) Use a compass. Draw a circle. Draw 2 different diameters.

   b) Use a protractor. Measure the angles
      at the centre of the circle.

   c) Find the sum of the angles.

   d) Repeat parts a to c for 3 different circles.
      What do you notice about the sum of the angles in each circle?

6. A glass has a circular base with radius 3.5 cm.
A rectangular tray has dimensions 40 cm by 25 cm.
How many glasses will fit on the tray?
What assumptions did you make?

7. **Assessment Focus** Your teacher will
give you a large copy of this logo.
Find the radius and diameter of each
circle in this logo. Show your work.

**This is the logo for the Aboriginal
Health Department of the
Vancouver Island Health Authority.**

8. **Take It Further** A circular area of grass needs watering.
A rotating sprinkler is to be placed at the centre of the circle.
Explain how you would locate the centre of the circle.
Include a diagram in your explanation.

Reflect

How are the diameter and radius of a circle related?
Include examples in your explanation.

4.2 Circumference of a Circle

Explore

You will need 3 circular objects of different sizes, string, and a ruler.

➤ Each of you chooses one of the objects.
Use string to measure the distance around it.
Measure the radius and diameter of the object.
Record these measures.

➤ Repeat the activity until each of you has measured all 3 objects.
Compare your results.
If your measures are the same, record them in a table.
If your measures for any object are different, measure again to check.
When you agree upon the measures, record them in the table.

Object	Distance Around (cm)	Radius (cm)	Diameter (cm)
Can			

➤ What patterns do you see in the table?
How is the diameter related
to the distance around?
How is the radius related
to the distance around?

➤ For each object, calculate:
  • distance around ÷ diameter
  • distance around ÷ radius
What do you notice?
Does the size of the circle affect
your answers? Explain.

Reflect & Share

Compare your results with those of another group.
Suppose you know the distance around a circle.
How can you find its diameter?

The distance around a circle is its **circumference**.

For any circle, the circumference, C, divided by the diameter, d, is approximately 3.

Circumference ÷ diameter $\doteq$ 3, or $\frac{C}{d} \doteq 3$

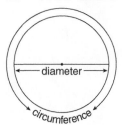

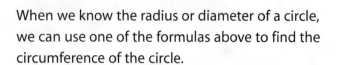

> The circumference of a circle is also the perimeter of the circle.

For any circle, the ratio $\frac{C}{d} = \pi$

The symbol π is a Greek letter that we read as "pi."

$\pi = 3.141\ 592\ 653\ 589\ldots$, or $\pi \doteq 3.14$

π is a decimal that never repeats and never terminates.
π cannot be written as a fraction.
For this reason, we call π an **irrational number**.

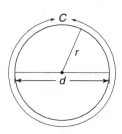

So, the circumference is π multiplied by d.
We write: $C = \pi d$

Since the diameter is twice the radius,
the circumference is also π multiplied by $2r$.
We write: $C = \pi \times 2r$, or $C = 2\pi r$

When we know the radius or diameter of a circle,
we can use one of the formulas above to find the
circumference of the circle.

The face of a toonie has radius 1.4 cm.

- To find the diameter of the face:
  The diameter $d = 2r$, where r is the radius
  Substitute: $r = 1.4$
  $$d = 2 \times 1.4$$
  $$= 2.8$$
  The diameter is 2.8 cm.

1.4 cm

> The circumference is a length, so its units are units of length such as centimetres, metres, or millimetres.

- To find the circumference of the face:

  $C = \pi d$ OR $C = 2\pi r$

  Substitute: $d = 2.8$ Substitute: $r = 1.4$

  $C = \pi \times 2.8$ $C = 2 \times \pi \times 1.4$

  $\doteq 8.796$ $\doteq 8.796$

  $\doteq 8.8$ $\doteq 8.8$

  The circumference is 8.8 cm, to one decimal place.

> Use the π key on your calculator. If the calculator does not have a π key, use 3.14 instead.

- We can estimate to check if the answer is reasonable.

  The circumference is approximately 3 times the diameter:

  $3 \times 2.8 \text{ cm} \doteq 3 \times 3 \text{ cm}$

  $= 9 \text{ cm}$

  The circumference is approximately 9 cm.

  The calculated answer is 8.8 cm, so this answer is reasonable.

When we know the circumference, we can use a formula to find the diameter.
Use the formula $C = \pi d$.
To isolate d, divide each side by π.

$$\frac{C}{\pi} = \frac{\pi d}{\pi}$$

$$\frac{C}{\pi} = d$$

So, $d = \frac{C}{\pi}$

Example

An above-ground circular swimming pool has circumference 12 m.
Calculate the diameter and radius of the pool.
Give the answers to two decimal places.
Estimate to check the answers are reasonable.

A Solution

The diameter is: $d = \frac{C}{\pi}$

Substitute: $C = 12$

$d = \frac{12}{\pi}$ Use a calculator.

$= 3.8197...$ Do not clear your calculator.

The radius is $\frac{1}{2}$ the diameter, or $r = d \div 2$.

Divide the number in the calculator display by 2.

$r \doteq 1.9099$

The diameter is 3.82 m to two decimal places.
The radius is 1.91 m to two decimal places.

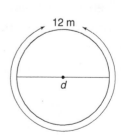

Since the circumference is approximately 3 times the diameter,
the diameter is about $\frac{1}{3}$ the circumference.
One-third of 12 m is 4 m. So, the diameter is about 4 m.
The radius is $\frac{1}{2}$ the diameter. One-half of 4 m is 2 m.
So, the radius of the pool is about 2 m.
Since the calculated answers are close to the estimates, the answers are reasonable.

Practice

1. Calculate the circumference of each circle.
   Give the answers to two decimal places.
   Estimate to check the answers are reasonable.

   a)

   ← 10 cm →

   b)

   7 cm

   c)

   ← 15 m →

2. Calculate the diameter and radius of each circle.
   Give the answers to two decimal places.
   Estimate to check the answers are reasonable.

   a)
   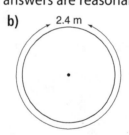
   24 cm

   b)
   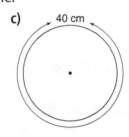
   2.4 m

   c)
   40 cm

3. When you estimate to check the circumference, you use 3 instead of π.
   Is the estimated circumference greater than or less than
   the actual circumference?
   Why do you think so?

4. A circular garden has diameter 2.4 m.
   a) The garden is to be enclosed with plastic edging.
      How much edging is needed?
   b) The edging costs $4.53/m.
      What is the cost to edge the garden?

5. **a)** Suppose you double the diameter of a circle.
What happens to the circumference?
b) Suppose you triple the diameter of a circle.
What happens to the circumference?
Show your work.

6. A carpenter is making a circular tabletop
with circumference 4.5 m.
What is the radius of the tabletop in centimetres?

Recall: 1 m = 100 cm

7. Can you draw a circle with circumference 33 cm?
If you can, draw the circle and explain how you know
its circumference is correct.
If you cannot, explain why it is not possible.

8. **Assessment Focus** A bicycle tire has a spot of wet paint on it.
The radius of the tire is 46 cm.
Every time the wheel turns, the paint marks the ground.
a) What pattern will the paint make on the ground as the bicycle moves?
b) How far will the bicycle have travelled between
two consecutive paint marks on the ground?
c) Assume the paint continues to mark the ground.
How many times will the paint mark the ground
when the bicycle travels 1 km?
Show your work.

9. **Take It Further** Suppose a metal ring could be placed
around Earth at the equator.
a) The radius of Earth is 6378.1 km. How long is the metal ring?
b) Suppose the length of the metal ring is increased by 1 km.
Would you be able to crawl under the ring, walk under the ring,
or drive a school bus under the ring?
Explain how you know.

Reflect

What is π?
How is it related to the circumference, diameter,
and radius of a circle?

Mid-Unit Review

LESSON

4.1

1. a) Use a compass.
Draw a circle with radius 3 cm.

b) Do not use a compass.
Draw a circle with radius 7 cm.
The circle should have the same
centre as the circle in part a.

2. Two circles have the same centre.
Their radii are 5 cm and 10 cm.
Another circle lies between these
circles. Give two possible
diameters for this circle.

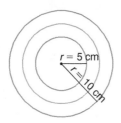

3. Find the radius of a circle with
each diameter.

a) 7.8 cm   **b)** 8.2 cm

c) 10 cm   **d)** 25 cm

4. Is it possible to draw two different
circles with the same radius and
diameter? Why or why not?

4.2

5. Calculate the circumference of
each circle. Give the answers to two
decimal places. Estimate to check
your answers are reasonable.

a)   **b)**

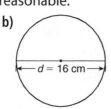

$r = 6$ cm $d = 16$ cm

6. a) Calculate the circumference of
each object.

   i) A wheelchair wheel with
diameter 66 cm

  ii) A tire with radius 37 cm

 iii) A hula-hoop with
diameter 60 cm

b) Which object has the greatest
circumference? How could you
tell without calculating the
circumference of each object?

7. Suppose the circumference
of a circular pond is 76.6 m.
What is its diameter?

8. Find the radius of a circle with
each circumference. Give your
answers to one decimal place.

a) 256 cm   **b)** 113 cm   **c)** 45 cm

9. An auger is used to drill a hole in
the ice, for ice fishing. The diameter
of the hole is 25 cm. What is the
circumference of the hole?

4.3 Area of a Parallelogram

Which of these shapes are parallelograms?
How do you know?

How are Shapes C and D alike?
How are they different?

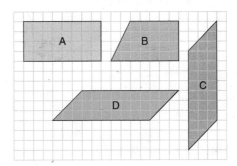

Explore

You will need scissors and 1-cm grid paper.

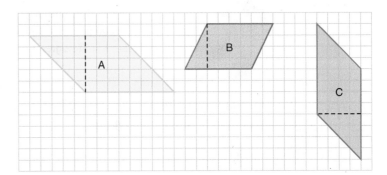

➤ Copy Parallelogram A
on grid paper.
Estimate, then find, the
area of the parallelogram.

➤ Cut out the parallelogram.
Then, cut along the broken
line segment.

➤ Arrange the two pieces to form a rectangle.
What is the area of the rectangle?
How does the area of the rectangle compare to the area
of the parallelogram?

➤ Repeat the activity for Parallelograms B and C.

Reflect & Share

Share your work with another pair of classmates.
Can every parallelogram be changed into a rectangle
by cutting and moving one piece? Explain.
Work together to write a rule for finding the area of a parallelogram.

To estimate the area of this parallelogram, count the whole squares and the part squares that are one-half or greater.

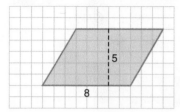

There are:
- 33 whole squares
- 8 part squares that are one-half or greater

The area of this parallelogram is about 41 square units.

Any side of a parallelogram is a **base** of the parallelogram. The **height** of a parallelogram is the length of a line segment that joins parallel sides and is perpendicular to the base.

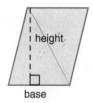

Recall that both a rectangle and a square are parallelograms.

Any parallelogram that is not a rectangle can be "cut" and rearranged to form a rectangle. Here is one way to do this.

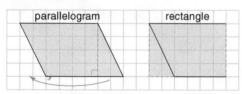

The parallelogram and the rectangle have the same area.
The area of a parallelogram is equal to the area of a rectangle with the same height and base.
To find the area of a parallelogram, multiply the base by the height.

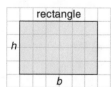

b represents the base.
h represents the height.

Area of rectangle:
$A = bh$

Area of parallelogram:
$A = bh$

Example

Calculate the area of each parallelogram.

a)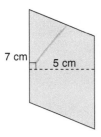

b)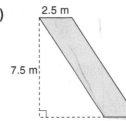

> The height can be drawn outside the parallelogram.

A Solution

The area of a parallelogram is given by the formula $A = bh$.

a) $A = bh$

Substitute: $b = 7$ and $h = 5$

$A = 7 \times 5$

$\quad = 35$

The area of the parallelogram is 35 cm².

b) $A = bh$

Substitute: $b = 2.5$ and $h = 7.5$

$A = 2.5 \times 7.5$

$\quad = 18.75$

The area of the parallelogram is 18.75 m².

Practice

1. i) Copy each parallelogram on 1-cm grid paper.

ii) Show how the parallelogram can be rearranged to form a rectangle.

iii) Estimate, then find, the area of each parallelogram.

a)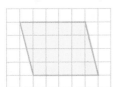

b)

c)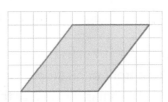

2. Find the area of each parallelogram.

a)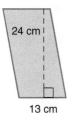

b)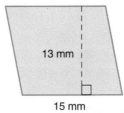

c)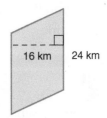

3. a) On 1-cm grid paper, draw 3 different parallelograms with base 3 cm and height 7 cm.

b) Find the area of each parallelogram you drew in part a. What do you notice?

4. Repeat question 3. This time, you choose the base and height.
Are your conclusions the same as in question 3? Why or why not?

5. Copy this parallelogram on 1-cm grid paper.
    a) Show how this parallelogram could be rearranged
        to form a rectangle.
    b) Find the area of the parallelogram.

6. Use the given area to find the base or the height of each parallelogram.
    a) Area = 60 m²   **b)** Area = 6 mm²   **c)** Area = 30 cm²

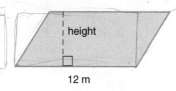
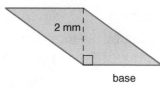
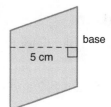

7. On 1-cm grid paper, draw as many different
parallelograms as you can with each area.
    a) 10 cm²   **b)** 18 cm²   **c)** 28 cm²

8. A student says the area of this parallelogram is 20 cm².
Explain the student's error.

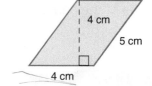

9. **Assessment Focus** Sasha is buying paint
for a design on a wall. Here is part of the design.
Sasha says Shape B will need
more paint than Shape A.
Do you agree? Why or why not?

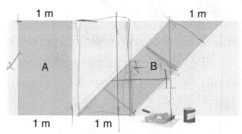

10. **Take It Further** A restaurant owner built a patio
in front of his store to attract more customers.
    a) What is the area of the patio?
    b) What is the total area of the patio and gardens?
    c) How can you find the area of the gardens?
Show your work.

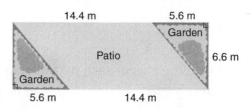

Reflect

How can you use what you know about rectangles
to help you find the area of a parallelogram?
Use an example to explain.

4.4 Area of a Triangle

Focus Develop and use a formula to find the area of a triangle.

Explore

You will need a geoboard, geobands, and dot paper.

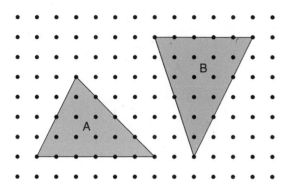

➤ Make Triangle A on a geoboard.
Add a second geoband to Triangle A to make
a parallelogram with the same base and height.
This is called a *related* parallelogram.
Make as many different parallelograms as you can.
How does the area of the parallelogram
compare to the area of Triangle A each time?
Record your work on dot paper.

➤ Repeat the activity with Triangle B.

➤ What is the area of Triangle A? Triangle B?
What strategy did you use to find the areas?

Reflect & Share

Share the different parallelograms you made with another pair of classmates.
Discuss the strategies you used to find the area of each triangle.
How did you use what you know about a parallelogram
to find the area of a triangle?
Work together to write a rule for finding the area of a triangle.

When we draw a diagonal in a parallelogram,
we make two congruent triangles.
Congruent triangles have the same area.
The area of the two congruent triangles
is equal to the area of the parallelogram
that contains them.
So, the area of one triangle is $\frac{1}{2}$ the area
of the parallelogram.

To find the area of this triangle:

Complete a parallelogram on
one side of the triangle.
The area of the parallelogram is:
$A =$ base $\times$ height, or $A = bh$
So, $A = 6 \times 5$
    $= 30$
The area of the parallelogram is 30 cm².
So, the area of the triangle is: $\frac{1}{2}$ of 30 cm² = 15 cm²

We can write a formula for the area of a triangle.
The area of a parallelogram is:
$A =$ base $\times$ height
So, the area of a triangle is:
$A =$ one-half of base $\times$ height
$A = bh \div 2$, which can be written as $A = \frac{bh}{2}$

Example

Find the area of each triangle.

a)
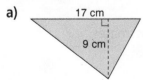

17 cm
9 cm

b)
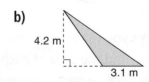

4.2 m
3.1 m

For an obtuse triangle,
the height might be drawn
outside the triangle.

A Solution

a) $A = \frac{bh}{2}$

Substitute: $b = 17$ and $h = 9$

$A = \frac{17 \times 9}{2}$

$= \frac{153}{2}$

$= 76.5$

The area is 76.5 cm².

b) $A = \frac{bh}{2}$

Substitute: $b = 3.1$ and $h = 4.2$

$A = \frac{3.1 \times 4.2}{2}$

$= \frac{13.02}{2}$

$= 6.51$

The area is 6.51 m².

Practice

1. Copy each triangle on 1-cm grid paper. Draw a related parallelogram.

a)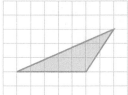

b)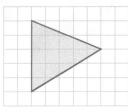

c)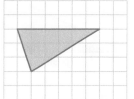

2. Each triangle is drawn on 1-cm grid paper.
Find the area of each triangle. Use a geoboard if you can.

a)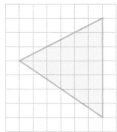

b)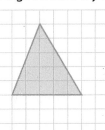

c)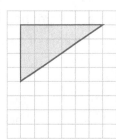

d)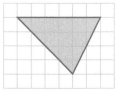

e)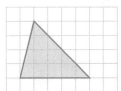

f)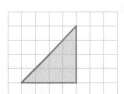

3. Draw two right triangles on 1-cm grid paper.
  a) Record the base and the height of each triangle.
  b) What do you notice about the height of a right triangle?
  c) Find the area of each triangle you drew.

4. a) Find the area of this triangle.

   b) Use 1-cm grid paper.
   How many different parallelograms can
   you draw that have the same base
   and the same height as this triangle?
   Sketch each parallelogram.

   c) Find the area of each parallelogram.
   What do you notice?

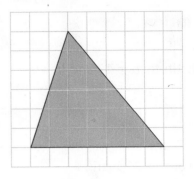

5. Use the given area to find the base or height of each triangle.
   How could you check your answers?

   a) Area = 18 cm²

   b) Area = 32 m²

   c) Area = 480 mm²

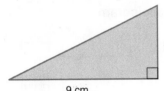
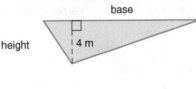
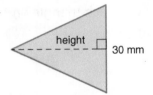

6. Use 1-cm grid paper.

   a) Draw 3 different triangles with each base and height.

   i) base: 1 cm; height: 12 cm

   ii) base: 2 cm; height: 6 cm

   iii) base: 3 cm; height: 4 cm

   b) Find the area of each triangle you drew in part a.
   What do you notice?

7. On 1-cm grid paper, draw two different triangles with each area below.
   Label the base and height each time.
   How do you know these measures are correct?

   a) 14 cm²   **b)** 10 cm²   **c)** 8 cm²

8. a) Draw any triangle on grid paper.
   What happens to the area of the triangle in each case?

   i) the base is doubled

   ii) both the height and the base are doubled

   iii) both the height and the base are tripled

   b) What could you do to the triangle you drew in part a to triple its area?
   Explain why this would triple the area.

9. Assessment Focus

This triangle is made from 4 congruent triangles.
Three triangles are to be painted blue.
The fourth triangle is not to be painted.

a) What is the area that is to be painted?
   Show your work.

b) The paint is sold in 1-L cans.
   One litre of paint covers 5.5 m².
   How many cans of paint are needed?
   What assumptions did you make?

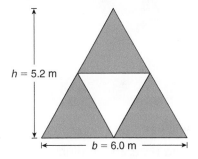

The height is approximate.

10. Look at the diagram to the right.

a) How many triangles do you see?
b) How are the triangles related?
c) How many parallelograms do you see?
d) Find the area of the large triangle.
e) Find the area of one medium-sized triangle.
f) Find the area of one small triangle.
g) Find the area of a parallelogram of your choice.

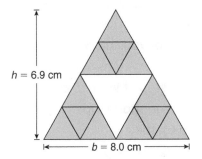

The height is approximate.

11. Take It Further

A local park has a pavilion to provide shelter.
The pavilion has a roof the shape of a rectangular pyramid.

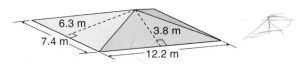

a) What is the total area of all four parts of the roof?
b) One sheet of plywood is 240 cm by 120 cm.
   What is the least number of sheets of plywood
   needed to cover the roof?
   Explain how you got your answer.

Reflect

What do you know about finding the area of a triangle?

Focus | Develop and use a formula to find the area of a circle.

Explore

You will need one set of fraction circles, masking tape, and a ruler.

➤ Each of you chooses one circle from the set of fraction circles. The circle you choose should have an even number of sectors, and at least 4 sectors.
➤ Each of you cuts 3 strips of masking tape:
  • 2 short strips
  • 1 strip at least 15 cm long
Use the short strips to fasten the long strip face up on the table.

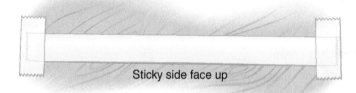

Sticky side face up

➤ Arrange all your circle sectors on the tape to approximate a parallelogram.
Trace your parallelogram, then use a ruler to make the horizontal sides straight.
Calculate the area of the parallelogram. Estimate the area of the circle.
How does the area of the parallelogram compare to the area of the circle?

Reflect & Share

Compare your measure of the area of the circle with the measures of your group members.
Which area do you think is closest to the area of the circle? Why?
How could you improve your estimate for the area?
Which circle measure best represents the height of the parallelogram?
The base? Work together to write a formula for the area of a circle.

Suppose a circle was cut into 8 congruent sectors.
The 8 sectors were then arranged to approximate a parallelogram.

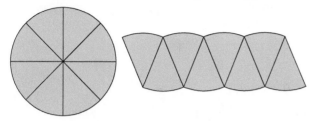

The more congruent sectors we use, the closer the area
of the parallelogram is to the area of the circle.
Here is a circle cut into 24 congruent sectors.
The 24 sectors were then arranged to approximate a parallelogram.

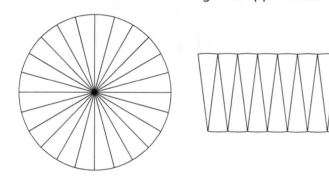

The greater the number of sectors, the more the shape looks like a rectangle.
The sum of the two longer sides of
the rectangle is equal to the circumference, C.
So, each longer side, or the base of the rectangle,
is one-half the circumference of the circle, or $\frac{C}{2}$.

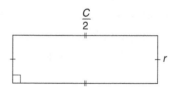

But $C = 2\pi r$
So, the base of the rectangle $= \frac{2\pi r}{2}$
$$= \pi r$$
Each of the two shorter sides is equal to the radius, r.

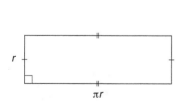
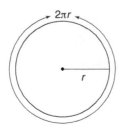

The area of a rectangle is: base × height
The base is πr. The height is r.
So, the area of the rectangle is: $\pi r \times r = \pi r^2$

Since the rectangle is made from all sectors of the circle,
the rectangle and the circle have the same area.
So, the area, A, of the circle with radius r is $A = \pi r^2$.

We can use this formula to find the area of any circle
when we know its radius.

When a number or variable
is multiplied by itself we
write: $7 \times 7 = 7^2$
         $r \times r = r^2$

$A = \pi r^2$

Example

The face of a dime has diameter 1.8 cm.
a) Calculate the area.
   Give the answer to two decimal places.
b) Estimate to check the answer is reasonable.

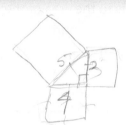

A Solution

The diameter of the face of a dime is 1.8 cm.
So, its radius is: $\frac{1.8 \text{ cm}}{2} = 0.9$ cm
a) Use the formula: $A = \pi r^2$
   Substitute: $r = 0.9$
   $A = \pi \times 0.9^2$
   Use a calculator.
   $A \doteq 2.544\ 69$
   The area of the face of the dime
   is 2.54 cm² to two decimal places.

1.8 cm

If your calculator does
not have an x² key, key in
0.9 × 0.9 instead.

b) Recall that $\pi \doteq 3$.
   So, the area of the face of the dime is about $3r^2$.
           $r \doteq 1$
       So, $r^2 = 1$
   and $3r^2 = 3 \times 1$
          $= 3$
   The area of the face of the dime is approximately 3 cm².
   Since the calculated area, 2.54 cm², is close to 3 cm²,
   the answer is reasonable.

1. Calculate the area of each circle.

Estimate to check your answers are reasonable.

a)
2 cm

b)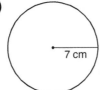
7 cm

c)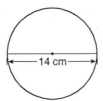
14 cm

d)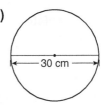
30 cm

2. Calculate the area of each circle. Give your answers to two decimal places.

Estimate to check your answers are reasonable.

a)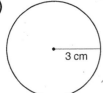
3 cm

b)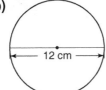
12 cm

c)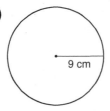
9 cm

d)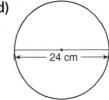
24 cm

3. Use the results of questions 1 and 2. What happens to the area in each case?
   a) You double the radius of a circle.
   b) You triple the radius of a circle.
   c) You quadruple the radius of a circle.
   Justify your answers.

4. **Assessment Focus** Use 1-cm grid paper.
   Draw a circle with radius 5 cm.
   Draw a square outside the circle that just encloses the circle.
   Draw a square inside the circle
   so that its vertices lie on the circle.
   Measure the sides of the squares.
   a) How can you use the areas of the two
   squares to estimate the area of the circle?
   b) Check your estimate in part a by calculating the area of the circle.
   c) Repeat the activity for circles with different radii.
   Record your results. Show your work.

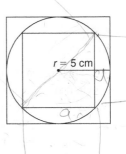
r = 5 cm

5. In the biathlon, athletes shoot at targets. Find the area of each target.
   a) The target for the athlete who is standing is a circle with diameter 11.5 cm.
   b) The target for the athlete who is lying down is a circle with diameter 4.5 cm.
   Give the answers to the nearest square centimetre.

6. In curling, the target area is a bull's eye
with 4 concentric circles.

   a) Calculate the area of the smallest circle.

   b) When a smaller circle overlaps a larger circle,
     a ring is formed.
     Calculate the area of each ring on the target area.

   Give your answers to 4 decimal places.

> Concentric circles have
> the same centre.

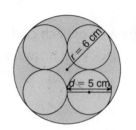

Radius 0.15 m
          0.60 m
          1.20 m
          1.80 m

 7. Take It Further

   A circle with radius 6 cm contains 4 small circles.
   Each small circle has diameter 5 cm.
   Each small circle touches two other
   small circles and the large circle.

   a) Find the area of the large circle.

   b) Find the area of one small circle.

   c) Find the area of the region that is shaded yellow.

$r = 6$ cm
$d = 5$ cm

8. Take It Further A large pizza has diameter 35 cm.
   Two large pizzas cost $19.99.
   A medium pizza has diameter 30 cm.
   Three medium pizzas cost $24.99.
   Which is the better deal: 2 large pizzas or 3 medium pizzas?
   Justify your answer.

Math Link

Agriculture: Crop Circles
In Red Deer, Alberta, on September 17, 2001,
a crop circle formation was discovered that
contained 7 circles. The circle shown has
diameter about 10 m. This circle destroyed
some wheat crop. What area of wheat crop
was lost in this crop circle?

Reflect

You have learned two formulas for measurements of a circle.
How do you remember which formula to use for the area of a circle?

Packing Circles

These circles are packed in a square.
In this game, you will pack circles in other shapes.

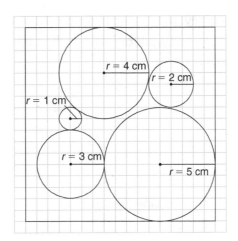

YOU WILL NEED

2 sheets of circles
scissors
ruler
compass
calculator

NUMBER OF PLAYERS

2

GOAL OF THE GAME

Construct the circle,
triangle, and parallelogram
with the lesser area.

> What strategies did you use
> to pack your circles to
> construct the shape with
> the lesser area?

HOW TO PLAY THE GAME:

1. Each player cuts out one sheet of circles.

2. Each player arranges his 5 circles so they
   are packed tightly together.

3. Use a compass. Draw a circle that encloses these circles.

4. Find the area of the enclosing circle.
   The player whose circle has the lesser area scores 2 points.

5. Pack the circles again.
   This time draw the parallelogram that encloses the circles.
   Find the area of the parallelogram.
   The player whose parallelogram has
   the lesser area scores 2 points.

6. Repeat *Step 5*. This time use a triangle to enclose the circles.

7. The player with the higher score wins.

Reading for Accuracy—Checking Your Work

Do you make careless mistakes on homework assignments?
On quizzes or tests?

To help improve your accuracy and test scores, you should
get in the habit of checking your work before handing it in.

The three most common types of mistakes are:
- copying errors
- notation errors
- calculation errors

Copying Errors

A copying error occurs when you copy the question
or numbers in the question incorrectly.

Find the copying error in this solution.
How can you tell that the solution is not reasonable? Explain.

1. Find 18% of 36.

$$18\% \text{ of } 36$$

$$36 \times .18 \longrightarrow \begin{array}{r} 36 \\ \times\ 1.8 \\ \hline 64.8 \end{array}$$

$$18\% \text{ of } 36 \text{ is } 64.8$$

Notation Errors

A notation error occurs when you use a math symbol incorrectly.

Find the notation error in this solution.

2. Evaluate: $(-8) + (+3) - (+2)$

Evaluate $(-8) + (+3) - (+2)$

$(-8) + (+3) - (+2) = (-5) - (+2)$

$\qquad\qquad\qquad\quad = (+5) + (+2)$

$\qquad\qquad\qquad\quad = (+7)$

Calculation Errors

A calculation error occurs when you make a mistake in your calculations.

Find the calculation error in this solution.
How can you tell that the solution is not reasonable? Explain.

3. A circular mat has diameter 60 cm.
What is the area of the mat?

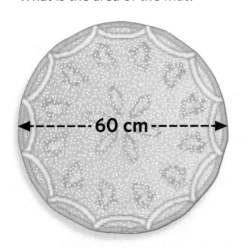

60 cm

The diameter of the mat is 60 cm.
So, its radius is: 60 cm/2 = 30 cm
Use the formula $A = \pi r^2$
Substitute: $r = 30$
$A = \pi \times 30^2$
$\quad = \pi \times 9000$
$\quad \doteq 28\ 274$
The area of the mat is about 28 274 cm².

4. Correct the error you found in each solution to find the correct answer.
Show your work.

Interpreting Circle Graphs

Focus | Interpret circle graphs to solve problems.

We can apply what we have learned about circles to interpret a new type of graph.

Explore

Sixty Grade 7 students at l'école Orléans were surveyed to find out their favourite after-school activity. The results are shown on the circle graph.

Which activity is most popular? Least popular? How do you know this from looking at the graph? How many students prefer each type of after-school activity? Which activity is the favourite for about $\frac{1}{3}$ of the students? Why do you think so? Write 3 more things you know from looking at the graph.

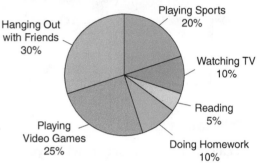

After-School Activities

Hanging Out with Friends 30%

Playing Sports 20%

Watching TV 10%

Reading 5%

Doing Homework 10%

Playing Video Games 25%

Reflect & Share

Compare your answers with those of another pair of classmates. What do you notice about the sum of the percents? Explain.

Connect

In a **circle graph**, data are shown as parts of one whole. Each **sector** of a circle graph represents a percent of the whole circle. The whole circle represents 100%.

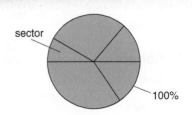

sector

100%

A circle graph has a title.

Each sector is labelled with a category and a percent.

A circle graph compares the number in each category to the total number.

That is, a fraction of the circle represents the same fraction of the total.

Sometimes, a circle graph has a **legend** that shows what category each sector represents.

In this case, only the percents are shown on the graph.

Favourite Sports of Grade 7 Students

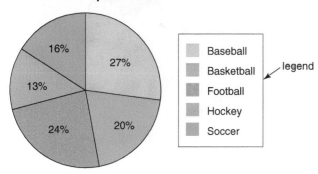

Example

This graph shows Nathan's typical day.

a) Which activity does Nathan do about $\frac{1}{4}$ of the time?

b) About how many hours does Nathan spend on each activity?

   Check that the answers are reasonable.

Nathan's Typical Day

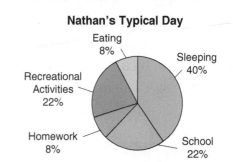

A Solution

a) Each of the sectors for "School" and "Recreational Activities" is about $\frac{1}{4}$ of the graph. 22% is close to 25%, which is $\frac{1}{4}$.

   So, Nathan is in school about $\frac{1}{4}$ of the day.

   He also participates in recreational activities about $\frac{1}{4}$ of the day.

b) ➤ From the circle graph, Nathan spends 40% of his day sleeping.

   There are 24 h in a day.

   Find 40% of 24.

   $40\% = \frac{40}{100} = 0.4$

   Multiply: $0.4 \times 24 = 9.6$ **9.6 is closer to 10 than to 9.**

   Nathan spends about 10 h sleeping.

➤ Nathan spends 22% of his day in school.

Find 22% of 24.

$22\% = \frac{22}{100} = 0.22$

Multiply: $0.22 \times 24 = 5.28$

> **5.28 is closer to 5 than to 6.**

Nathan spends about 5 h in school.

Nathan also spends about 5 h doing recreational activities.

➤ Nathan spends 8% of his day doing homework.

Find 8% of 24.

$8\% = \frac{8}{100} = 0.08$

Multiply: 0.08×24

Multiply as you would whole numbers.

$$\begin{array}{r} 24 \\ \times\ 8 \\ \hline 192 \end{array}$$

Estimate to place the decimal point.

$0.1 \times 24 = 2.4$

So, $0.08 \times 24 = 1.92$

> **1.92 is closer to 2 than to 1.**

Nathan spends about 2 h doing homework.

Nathan also spends about 2 h eating.

The total number of hours spent on all activities
should be 24, the number of hours in a day:

$9.6 + 5.28 + 5.28 + 1.92 + 1.92 = 24$

So, the answers are reasonable.

> **Add the exact times, *not* the approximate times.**

Practice

1. This circle graph shows the most popular
   activities in a First Nations school.
   There are 500 students in the school.
   All students voted.
   a) Which activity did about $\frac{1}{4}$ of
      the students choose?
      How can you tell by looking at the graph?
   b) Which activity is the most popular? The least popular?
   c) Find the number of students who chose each activity.
   d) How can you check your answers to part c?

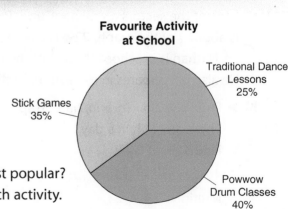

**Favourite Activity
at School**

Traditional Dance
Lessons
25%

Stick Games
35%

Powwow
Drum Classes
40%

2. This circle graph shows the ages of viewers of a TV show.

One week, approximately 250 000 viewers tuned in.

a) Which two age groups together make up $\frac{1}{2}$ of the viewers?

b) How many viewers were in each age group?

   i) 13 to 19   **ii)** 20 to 29   **iii)** 40 and over

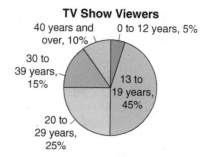

TV Show Viewers

3. This graph shows the world's gold production for a particular year.
In this year, the world's gold production was approximately 2300 t.
About how much gold would have been produced in each country?

a) Canada   **b)** South Africa

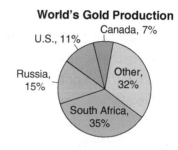

World's Gold Production

4. The school library budget to buy new books is $5000.
The librarian has this circle graph to show the types of books students borrowed in one year.

a) How much money should be spent on each type of book? How do you know?

b) Explain how you can check your answers in part a.

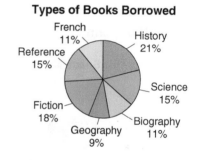

Types of Books Borrowed

5. **Assessment Focus** This circle graph shows the populations of the 4 Western Canadian provinces in 2005.
The percent for Saskatchewan is not shown.

a) What percent of the population lived in Saskatchewan? How do you know?

b) List the provinces in order from least to greatest population.
How did the circle graph help you do this?

c) In 2005, the total population of the Western provinces was about 9 683 000 people.
Calculate the population of each province, to the nearest thousand.

d) What else do you know from looking at the circle graph? Write as much as you can.

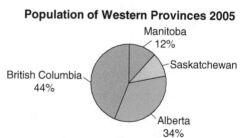

Population of Western Provinces 2005

6. Gaston collected data about the favourite season of his classmates.

Favourite Season

Classmates' Favourite Season

Season	Autumn	Winter	Spring	Summer
Number of Students	7	3	5	10

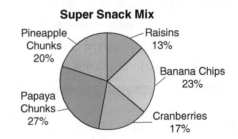

He recorded the results in a circle graph.
The graph is not complete.

a) How many students were surveyed?

b) Write the number of students who chose each season as a fraction of the total number of students, then as a percent.

c) Explain how you can check your answers to part b.

d) Sketch the graph. Label each sector with its name and percent. How did you do this?

7. These circle graphs show the percent of ingredients in two 150-g samples of different snack mixes.

Morning Snack Mix

Peanuts 27%
Sunflower Seeds 20%
Raisins 17%
Almonds 36%

Super Snack Mix

Pineapple Chunks 20%
Raisins 13%
Banana Chips 23%
Papaya Chunks 27%
Cranberries 17%

a) For each snack mix, calculate the mass, in grams, of each ingredient.

b) About what mass of raisins would you expect to find in a 300-g sample of each mix? What assumptions did you make?

Reflect

Search newspapers, magazines, and the Internet to find examples of circle graphs.
Cut out or print the graphs.
How are they the same? How are they different?
Why were circle graphs used to display these data?

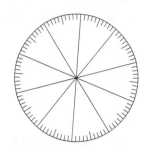

Focus Construct circle graphs to display data.

This is a **percent circle**.
The circle is divided into 100 congruent parts.
Each part is 1% of the whole circle.
You can draw a circle graph on a percent circle.

Explore

Your teacher will give you a percent circle.
Students in a Grade 7 class were asked
how many siblings they have.
Here are the results.

0 Siblings	1 Sibling	2 Siblings	More than 2 Siblings
3	13	8	1

Write each number of students as a fraction of the total number.
Then write the fraction as a percent.

Use the percent circle.
Draw a circle graph to display the data.
Write 2 questions you can answer by looking at the graph.

Reflect & Share

Trade questions with another pair of classmates.
Use your graph to answer your classmates' questions.
Compare graphs. If they are different, try to find out why.
How did you use fractions and percents to draw a circle graph?

Connect

Recall that a circle graph shows how parts of a set of data
compare with the whole set.
Each piece of data is written as a fraction of the whole.
Each fraction is then written as a percent.
Sectors of a percent circle are coloured to represent these percents.
The sum of the **central angles** is 360°.
A central angle is also called a **sector angle**.

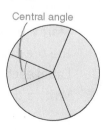

Central angle

Example

All the students in two Grade 7 classes
were asked how they get to school each day.
Here are the results: 9 rode their bikes, 11 walked,
17 rode the bus, and 13 were driven by car.
Construct a circle graph to illustrate these data.

A Solution

➤ For each type of transport:

Write the number of students as a fraction of 50,
the total number of students.

Then write each fraction as a decimal and as a percent.

Bike: $\frac{9}{50} = \frac{18}{100} = 0.18 = 18\%$ Walk: $\frac{11}{50} = \frac{22}{100} = 0.22 = 22\%$

Bus: $\frac{17}{50} = \frac{34}{100} = 0.34 = 34\%$ Car: $\frac{13}{50} = \frac{26}{100} = 0.26 = 26\%$

The circle represents all the types of transport.
To check, add the percents.
The sum should be 100%.

$18\% + 22\% + 34\% + 26\% = 100\%$

> **Another Strategy**
> We could use a percent
> circle to graph these data.

➤ To find the sector angle for each type of transport,
multiply each decimal by 360°.
Write each angle to the nearest degree, when necessary.

Bike 18%: $0.18 \times 360° = 64.8° \doteq 65°$

Walk 22%: $0.22 \times 360° = 79.2° \doteq 79°$

Bus 34%: $0.34 \times 360° = 122.4° \doteq 122°$

Car 26%: $0.26 \times 360° = 93.6° \doteq 94°$

Check:
$64.8° + 79.2° + 122.4° + 93.6° = 360°$

➤ Construct a circle.
Use a protractor to construct
each sector angle.
Start with the smallest angle.
Draw a radius. Measure 65°.

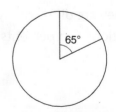

Start the next sector where
the previous sector finished.
Label each sector with its name and percent.
Write a title for the graph.

How Students Get to School

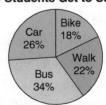

1. The table shows the number of Grade 7 students with each eye colour at Northern Public School.

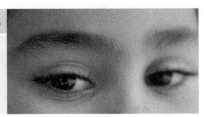

Eye Colour	Number of Students
Blue	12
Brown	24
Green	8
Grey	6

   a) Find the total number of students.
   b) Write the number of students with each eye colour as a fraction of the total number of students.
   c) Write each fraction as a percent.
   d) Draw a circle graph to represent these data.

2. In a telephone survey, 400 people voted for their favourite radio station.

Radio Station	Votes
MAJIC99	88
EASY2	?
ROCK1	120
HITS2	100

   a) How many people chose EASY2?
   b) Write the number of people who voted for each station as a fraction of the total number who voted. Then write each fraction as a percent.
   c) Draw a circle graph to display the results of the survey.

3. **Assessment Focus** This table shows the method of transport used by U.S. residents entering Canada in one year.

United States Residents Entering Canada

Method of Transport	Number
Automobile	32 000 000
Plane	4 000 000
Train	400 000
Bus	1 600 000
Boat	1 200 000
Other	800 000

   a) How many U.S. residents visited Canada that year?
   b) What fraction of U.S. residents entered Canada by boat?
   c) What percent of U.S. residents entered Canada by plane?
   d) Display the data in a circle graph.
   e) What else do you know from the table or circle graph? Write as much as you can.

4. Can the data in each table below be displayed in a circle graph? Explain.

a)

Educational Attainment of Canadians	
0 to 8 years of elementary school	10%
Some secondary school	17%
Graduated from high school	20%
Some post-secondary education	9%
Post-secondary certificate or diploma	28%
University degree	16%

b)

Canadian Households with These Conveniences	
Automobile	64%
Cell phone	42%
Dishwasher	51%
Internet	42%

5. Take It Further This circle graph shows the percent of land occupied by each continent.
The area of North America is approximately 220 million km^2.
Use the percents in the circle graph.
Find the approximate area of each of the other continents,
to the nearest million square kilometres.

Area of Land

Australia (5%)
Europe (7%)
Antarctica (8%)
Asia (30%)
S. America (12%)
Africa (20%)
N. America (18%)

Reflect

When is it most appropriate to show data using a circle graph?
When is it not appropriate?

Using a Spreadsheet to Create Circle Graphs

Focus Display data on a circle graph using spreadsheets.

Spreadsheet software can be used to record, then graph, data. This table shows how Stacy budgets her money each month.

Stacy's Monthly Budget

Category	Amount ($)
Food	160
Clothing	47
Transportation	92
Entertainment	78
Savings	35
Rent	87
Other	28

Enter the data into rows and columns of a spreadsheet. Highlight the data. Do not include the column heads.

	A	B
1	**Category**	**Amount ($)**
2	Food	160
3	Clothing	47
4	Transportation	92
5	Entertainment	78
6	Savings	35
7	Rent	87
8	Other	28

Click the graph/chart icon. In most spreadsheet programs, circle graphs are called **pie charts**. Select *pie chart*. Investigate different ways of labelling the graph.
Your graph should look similar to one of the graphs on the following page.

Stacy's Monthly Budget

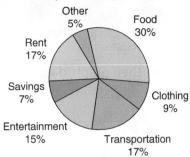

Other 5%
Food 30%
Rent 17%
Savings 7%
Clothing 9%
Entertainment 15%
Transportation 17%

Stacy's Monthly Budget

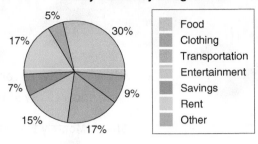

5%
30%
17%
7%
9%
15%
17%

- Food
- Clothing
- Transportation
- Entertainment
- Savings
- Rent
- Other

This circle graph shows
a legend at the right.
The legend shows what category
each sector represents.

These data are from *Statistics Canada*.

1. a) Use a spreadsheet.
   Create a circle graph to display
   these data.
   b) Write 3 questions about your graph.
   Answer your questions.
   c) Compare your questions
   with those of a classmate.
   What else do you know from
   the table or the graph?

Population by Province and Territory, October 2005

Region	Population
Newfoundland and Labrador	515 591
Prince Edward Island	138 278
Nova Scotia	938 116
New Brunswick	751 726
Quebec	7 616 645
Ontario	12 589 823
Manitoba	1 178 109
Saskatchewan	992 995
Alberta	3 281 296
British Columbia	4 271 210
Yukon Territories	31 235
Northwest Territories	42 965
Nunavut	30 133

Unit Review

What Do I Need to Know?

✓ Measurements in a Circle

The distance from the centre to a point on the circle
is the *radius*. The distance across the circle,
through the centre, is the *diameter*.
The distance around the circle is the *circumference*.

✓ Circle Relationships

In a circle, let the radius be *r*, the diameter *d*,
the circumference *C*, and the area *A*.

Then, $d = 2r$

$\frac{d}{2} = r$

$C = 2\pi r$, or $C = \pi d$

$A = \pi r^2$

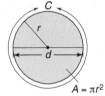

π is an irrational number that
is approximately 3.14.
The sum of the central angles
of a circle is 360°.

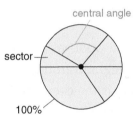

✓ Area Formulas

Parallelogram: $A = bh$
where *b* is the base and
h is the height

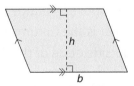

Triangle: $A = \frac{bh}{2}$ or
$\qquad A = bh \div 2$
where *b* is the base and *h* is the height

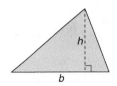

✓ Circle Graphs

In a circle graph, data are shown as parts of one whole.
The data are reported as a percent of the total, and the sum
of the percents is 100%. The sum of the sector angles is 360°.

What Should I Be Able to Do?

LESSON

4.1
1. Draw a large circle without using a compass.
Explain how to find the radius and diameter of the circle you have drawn.

2. Find the radius of a circle with each diameter.
   a) 12 cm   b) 20 cm   c) 7 cm

3. Find the diameter of a circle with each radius.
   a) 15 cm   b) 22 cm   c) 4.2 cm

4.2
4. The circumference of a large crater is about 219 m.
What is the radius of the crater?

5. A circular pool has a circular concrete patio around it.
   a) What is the circumference of the pool?
   b) What is the combined radius of the pool and patio?
   c) What is the circumference of the outside edge of the patio?

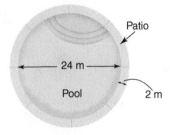

6. Mitra and Mel have different MP3 players.
The circular control dial on each player is a different size.
Calculate the circumference of the dial on each MP3 player.
   a) Mitra's dial: diameter 30 mm
   b) Mel's dial: radius 21 mm
   c) Whose dial has the greater circumference? Explain.

4.3
7. On 0.5-cm grid paper, draw 3 different parallelograms with area 24 cm². What is the base and height of each parallelogram?

4.3
4.4
8. a) The window below consists of 5 pieces of glass. Each piece that is a parallelogram has base 1.6 m. What is the area of one parallelogram?

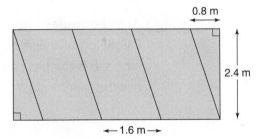

   b) The base of each triangle in the window above is 0.8 m.
   i) What is the area of one triangle?
   ii) What is the area of the window?
   Explain how you found the area.

9. On 0.5-cm grid paper, draw 3 different triangles with area 12 cm².

  a) What is the base and height of each triangle?

  b) How are the triangles related to the parallelograms in question 7?

10. Po Ling is planning to pour a concrete patio beside her house. It has the shape of a triangle. The contractor charges $125.00 for each square metre of patio.

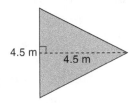

4.5 m
4.5 m

What will the contractor charge for the patio?

11. A goat is tied to an 8-m rope in a field.

  a) What area of the field can the goat graze?

  b) What is the circumference of the area in part a?

12. Choose a radius. Draw a circle. Suppose you divide the radius by 2.

  a) What happens to the circumference?

  b) Explain what happens to the area.

13. The diameter of a circular mirror is 28.5 cm. What is the area of the mirror? Give the answer to two decimal places.

14. Suppose you were to paint inside each shape below. Which shape would require the most paint? How did you find out?

  a)

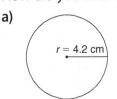

$r = 4.2$ cm

  b)

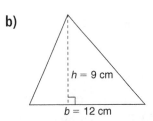

$h = 9$ cm
$b = 12$ cm

  c)

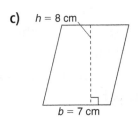

$h = 8$ cm
$b = 7$ cm

4.6 **15.** The results of the student council election are displayed on a circle graph. Five hundred students voted. The student with the most votes was named president.

a) Which student was named president? How do you know?

b) How many votes did each candidate receive?

c) Write 2 other things you know from the graph.

Student Council Election Results

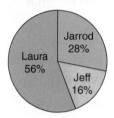

16. This circle graph shows the surface areas of the Great Lakes.

Areas of the Great Lakes

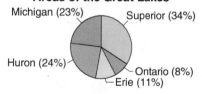

a) Which lake has a surface area about $\frac{1}{4}$ of the total area?

b) Explain why Lake Superior has that name.

c) The total area of the Great Lakes is about 244 000 km². Find the surface area of Lake Erie.

4.7 **17.** This table shows the approximate chemical and mineral composition of the human body.

Component	Percent
Water	62
Protein	17
Fat	15
Nitrogen	3
Calcium	2
Other	1

a) Draw a circle graph to display these data.

b) Jensen has mass 60 kg. About how many kilograms of Jensen's mass is water?

18. Here are the top 10 point scorers on the 2006 Canadian Women's Olympic Hockey Team. The table shows each player's province of birth.

Manitoba	Saskatchewan
Botterill	Wickenheiser

Quebec	Ontario
Ouellette	Apps
Goyette	Campbell
Vaillancourt	Hefford
	Piper
	Weatherston

a) What percent was born in each province?

b) Draw a circle graph to display the data in part a.

c) Why do you think more of these players come from Ontario than from any other province?

Practice Test

1. Draw a circle. Measure its radius.
   Calculate its diameter, circumference, and area.

2. The circular frame of this dream catcher has diameter 10 cm.
   a) How much wire is needed to make
      the outside frame?
   b) What is the area enclosed by the frame
      of this dream catcher?

3. A circle is divided into 8 sectors.
   What is the sum of the central angles
   of the circle? Justify your answer.

4. Find the area of each shape. Explain your strategy.

   a)
   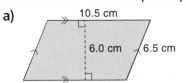

   b)
   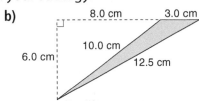

5. a) How many different triangles and parallelograms
      can you sketch with area 20 cm²?
      Have you sketched all possible shapes? Explain.
   b) Can you draw a circle with area 20 cm²?
      If your answer is yes, explain how you would do it.
      If your answer is no, explain why you cannot do it.

6. The table shows the type of land cover
   in Canada, as a percent of the total area.
   a) Draw a circle graph.
   b) Did you need to know the area of Canada
      to draw the circle graph? Explain.
   c) Write 3 things you know from looking
      at the graph.

Type of Land Cover in Canada	
Forest and taiga	45%
Tundra	23%
Wetlands	12%
Fresh water	8%
Cropland and rangeland	8%
Ice and snow	3%
Human use	1%

An anonymous donor gave a large sum of money to the
Parks and Recreation Department.
The money is to be used to build a large circular water park.
Your task is to design the water park.

The water park has radius 30 m.

The side length of each square on this grid represents 4 m.

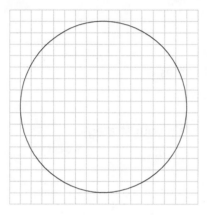

You must include the following features:

2 Wading Pools:

Each wading pool is triangular.
The pools do not have the same dimensions.
Each pool has area 24 m².

3 Geysers:

A geyser is circular.
Each geyser sprays water out of the ground,
and soaks a circular area with diameter 5 m or 10 m.

2 Wet Barricades:

A barricade has the shape of a parallelogram.
A row of nozzles in the barricade shoots water vertically.
The water falls within the area of the barricade.

4 Time-out Benches:

Each bench is shaped like a parallelogram.
It must be in the park.

At Least 1 Special Feature:

This feature will distinguish your park from other parks.
This feature can be a combination of any of the
shapes you learned in this unit.
Give the dimensions of each special feature.
Explain why you included each feature in the park.

Your teacher will give you a grid to draw your design.
You may use plastic shapes or cutouts to help you plan your park.
Complete the design.
Colour the design to show the different features.

Design your park so that a person can walk through
the middle of the park without getting wet.
What area of the park will get wet?

Check List

Your work
should show:

✓ the area of each
different shape you
used

✓ a diagram of your
design on grid paper

✓ an explanation of
how you created the
design

✓ how you calculated
the area of the park
that gets wet

Reflect on Your Learning

You have learned to measure different shapes.
When do you think you might use this knowledge outside the classroom?

Investigation

Digital Roots

Work with a partner.

The **digital root** of a number is the result of adding the digits of the number until a single-digit number is reached.

For example, the digital root of 27 is: $2 + 7 = 9$
To find the digital root of 168:
Add the digits: $1 + 6 + 8 = 15$
Since 15 is not a single-digit number, add the digits: $1 + 5 = 6$
Since 6 is a single-digit number, the digital root of 168 is 6.

A digital root can also be found for the product of a multiplication fact.
For the multiplication fact, 8×4:
$8 \times 4 = 32$
Add the digits in the product: $3 + 2 = 5$
Since 5 is a single-digit number, the digital root of 8×4 is 5.

You will explore the digital roots of the products in a multiplication table, then display the patterns you find.
As you complete the *Investigation*, include all your work in a report that you will hand in.

Materials:
- multiplication chart
- compass
- protractor
- ruler

Part 1

➤ Use a blank 12 × 12 multiplication chart.
Find each product.
Find the digital root of each product.
Record each digital root in the chart.
For example, for the product $4 \times 4 = 16$,
the digital root is $1 + 6 = 7$.

×	1	2	3	4	...
1					
2					
3					
4				7	
⋮					

➤ Describe the patterns in the completed chart.
Did you need to calculate the digital root of each product?
Did you use patterns to help you complete the table?
Justify the method you used to complete the chart.

➤ Look down each column. What does each column represent?

Part 2

➤ Use a compass to draw 12 circles.
Use a protractor to mark 9 equally spaced points
on each circle.
Label these points in order, clockwise, from 1 to 9.
Use the first circle.
Look at the first two digital roots in the 1st column
of your chart.
Find these numbers on the circle.
Use a ruler to join these numbers with a line segment.
Continue to draw line segments to join points that match
the digital roots in the 1st column.
What shape have you drawn?

➤ Repeat this activity for each remaining column.
Label each circle with the number at the top of the column.

➤ Which circles have the same shape?
Which circle has a unique shape?
What is unique about the shape?
Why do some columns have the same pattern of digital roots?
Explain.

Take It Further

➤ Investigate if similar patterns occur in each case:
  • Digital roots of larger 2-digit numbers, such as 85 to 99
  • Digital roots of 3-digit numbers, such as 255 to 269
  Write a report on what you find out.

Operations with Fractions

Many newspapers and magazines sell advertising space. Why would a company pay for an advertisement?

Selling advertising space is a good way to raise funds. Students at Anishwabe School plan to sell advertising space in their yearbook.

How can fractions be used in advertising space?

What You'll Learn

- Add and subtract fractions using models, pictures, and symbols.
- Add and subtract mixed numbers.
- Solve problems involving the addition and subtraction of fractions and mixed numbers.

Why It's Important

- You use fractions when you read gauges, shop, measure, and work with percents and decimals, and in sports, recipes, and business.

Key Words

- fraction strips
- simplest form
- related denominators
- unrelated denominators
- common denominator
- unit fraction

Let the yellow hexagon represent 1:

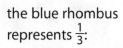

Then the red trapezoid represents $\frac{1}{2}$:

the blue rhombus represents $\frac{1}{3}$:

and the green triangle represents $\frac{1}{6}$:

Explore

Use Pattern Blocks.

Bakana trains for cross-country one hour a day. Here is her schedule:
Run for $\frac{1}{3}$ of the time, walk for $\frac{1}{6}$ of the time,
then run for the rest of the time.
How long does Bakana run altogether?
What fraction of the hour is this?

- Use fractions to write an addition equation to show how Bakana spent her hour.
- Bakana never runs for the whole hour.
  Write another possible schedule for Bakana.
  Write an addition equation for the schedule.
- Trade schedules with another pair of classmates.
  Write an addition equation for your classmates' schedule.

Reflect & Share

For the same schedule, compare equations with another pair of classmates.
Were the equations the same? How can you tell?
When are Pattern Blocks a good model for adding fractions?
When are Pattern Blocks not a good model?

There are many models that help us add fractions.

- We could use clocks to model halves, thirds, fourths, sixths, and twelfths.

 $\frac{1}{2} + \frac{1}{3} + \frac{1}{12} = \frac{11}{12}$

Circle models are useful when the fractions are less than 1.

- The example below uses fraction circles to add fractions.

Example

Zack and Ronny each bought a small pizza.
Zack ate $\frac{3}{4}$ of his pizza and Ronny ate $\frac{7}{8}$ of his.
How much pizza did Zack and Ronny eat together?

A Solution

Add: $\frac{3}{4} + \frac{7}{8}$

Use fraction circles.

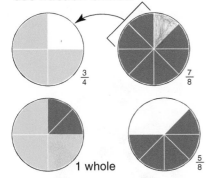

Use eighths to fill the circle for $\frac{3}{4}$.
Two-eighths fill the circle.

1 whole and 5 eighths equals $1\frac{5}{8}$.
So, $\frac{3}{4} + \frac{7}{8} = 1\frac{5}{8}$

Together, Zack and Ronny ate $1\frac{5}{8}$ pizzas.

Practice

Use Pattern Blocks or fraction circles.

1. Model each picture. Then, find each sum.

a)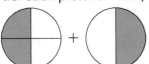

b)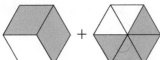

c)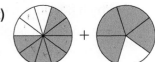

2. Use a model to show each sum. Sketch the model.
Write an addition equation for each picture.

a) $\frac{7}{8} + \frac{1}{2}$ b) $\frac{3}{10} + \frac{2}{5}$ c) $\frac{2}{3} + \frac{1}{2}$ d) $\frac{2}{3} + \frac{5}{6}$

e) $\frac{3}{6} + \frac{1}{12}$ f) $\frac{1}{4} + \frac{2}{8}$ g) $\frac{1}{3} + \frac{1}{2}$ h) $\frac{1}{2} + \frac{4}{10}$

3. Simon spends $\frac{1}{6}$ h practising the whistle flute each day.

He also spends $\frac{1}{3}$ h practising the drums.

How much time does Simon spend each day practising these instruments?

Show how you found your solution.

4. a) Add.

i) $\frac{1}{5} + \frac{1}{5}$ ii) $\frac{2}{3} + \frac{1}{3}$ iii) $\frac{4}{10} + \frac{3}{10}$ iv) $\frac{1}{6} + \frac{3}{6}$

b) Look at your work in part a. How did you find your solutions?
How else could you add fractions with like denominators?

5. Is each sum greater than 1 or less than 1? How can you tell?

a) $\frac{1}{4} + \frac{2}{4}$ b) $\frac{2}{5} + \frac{7}{5}$ c) $\frac{3}{4} + \frac{1}{4}$ d) $\frac{1}{10} + \frac{3}{10}$

6. Assessment Focus   Bella added 2 fractions. Their sum was $\frac{5}{6}$.
Which 2 fractions might Bella have added?
Find as many pairs of fractions as you can.
Show your work.

7. Asani's family had bannock with their dinner.
The bannock was cut into 8 equal pieces.
Asani ate 1 piece, her brother ate 2 pieces,
and her mother ate 3 pieces.

a) What fraction of the bannock did Asani eat?
Her brother? Her mother?

b) What fraction of the bannock was eaten?
What fraction was left?

Reflect

Which fractions can you add using Pattern Blocks? Fraction circles?
Give an example of fractions for which you cannot
use these models to add.

Using Other Models to Add Fractions

Focus Use fraction strips and number lines to add fractions.

We can use an area model to show fractions of one whole.

Explore

Your teacher will give you a copy of the map.
The map shows a section of land owned by 6 people.

- What fraction of land did each person own?
  What strategies did you use to find out?

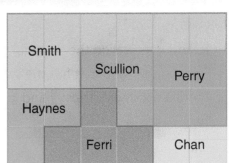

Three people sold land to the other 3 people.

- Use the clues below to draw the new map.
- Write addition equations, such as $\frac{1}{2} + \frac{1}{4} = \frac{3}{4}$,
  to keep track of the land sales.

CLUES

1. When all the sales were finished, four people owned all the land — Smith, Perry, Chan, and Haynes.
2. Smith now owns $\frac{1}{2}$ of the land.
3. Perry kept $\frac{1}{2}$ of her land, and sold the other half.
4. Chan bought land from two other people. He now owns $\frac{1}{4}$ of the land.
5. Haynes now owns the same amount of land as Perry started with.

Reflect & Share

Did you find any equivalent fractions? How do you know they are equivalent?
Which clues helped you most to draw the new map? Explain how they helped.

Connect

You can model fractions with strips of paper called **fraction strips**.

Here are more fraction strips and some equivalent fractions they show.

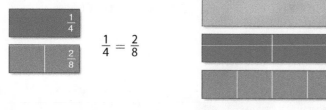

$$\frac{1}{4} = \frac{2}{8}$$

$$\frac{1}{2} = \frac{2}{4} = \frac{4}{8}$$

Recall that equivalent fractions show the same amount.

This strip represents 1 whole.

To add $\frac{1}{4} + \frac{1}{2}$, align the strips for $\frac{1}{4}$ and $\frac{1}{2}$.

Find a single strip that has the same length as the two strips.

There are 2 single strips: $\frac{6}{8}$ and $\frac{3}{4}$

So, $\frac{1}{4} + \frac{1}{2} = \frac{6}{8}$

And, $\frac{1}{4} + \frac{1}{2} = \frac{3}{4}$

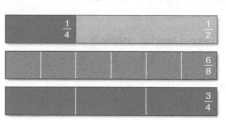

$\frac{3}{4}$ and $\frac{6}{8}$ are equivalent fractions.

The fraction $\frac{3}{4}$ is in simplest form.

A fraction is in **simplest form** when the numerator and denominator have no common factors other than 1.

When the sum is greater than 1, we could use fraction strips and a number line.

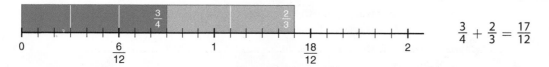

$$\frac{3}{4} + \frac{2}{3} = \frac{17}{12}$$

Example
Add. $\frac{1}{2} + \frac{4}{5}$

A Solution
$\frac{1}{2} + \frac{4}{5}$

Place both strips end-to-end on the halves line.

The right end of the $\frac{4}{5}$-strip does not line up with a fraction on the halves line.

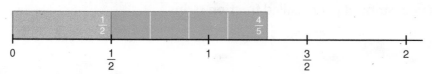

Place both strips on the fifths line.

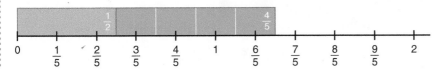

The right end of the $\frac{4}{5}$-strip does not line up with a fraction on the fifths line. Find a line on which to place both strips so the end of the $\frac{4}{5}$-strip lines up with a fraction.

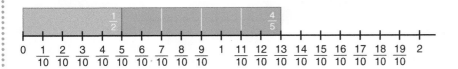

Another Strategy
We could add these fractions using fraction circles.

The end of the $\frac{4}{5}$-strip lines up with a fraction on the tenths line. The strips end at $\frac{13}{10}$. So, $\frac{1}{2} + \frac{4}{5} = \frac{13}{10}$

Practice

Use fraction strips and number lines.

1. Use the number lines below. List all fractions equivalent to:

    a) $\frac{1}{2}$ b) $\frac{1}{4}$ c) $\frac{2}{3}$

    Use a ruler to align the fractions if it helps.

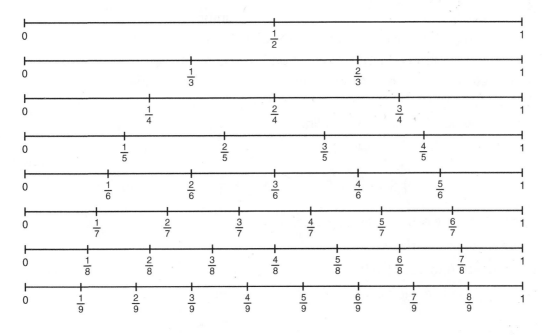

2. Write an addition equation for each picture.

a)

$\frac{3}{4}$ $\frac{7}{8}$

0 ——— $\frac{4}{8}$ ——— 1 ——— $\frac{12}{8}$ ——— 2

b)

$\frac{5}{6}$ $\frac{2}{3}$

0 ——— $\frac{3}{6}$ ——— 1 ——— $\frac{9}{6}$ ——— 2

c)

$\frac{3}{2}$ $\frac{3}{4}$

0 ——— $\frac{2}{4}$ ——— 1 ——— $\frac{6}{4}$ ——— 2 ——— $\frac{10}{4}$ ——— 3

3. Use your answers to question 2.

a) Look at the denominators in each part, and the number line
   you used to get the answer. What patterns do you see?

b) The denominators in each part of question 2 are **related denominators**.
   Why do you think they have this name?

4. Add.

a) $\frac{1}{3} + \frac{5}{6}$

b) $\frac{7}{12} + \frac{1}{3}$

c) $\frac{3}{5} + \frac{1}{10}$

d) $\frac{1}{6} + \frac{1}{12}$

5. Add.

a) $\frac{1}{3} + \frac{1}{2}$

b) $\frac{3}{4} + \frac{5}{6}$

c) $\frac{3}{5} + \frac{1}{2}$

d) $\frac{2}{3} + \frac{1}{5}$

6. Look at your answers to question 5.

a) Look at the denominators in each part, and the number line
   you used to get the answer. What patterns do you see?

b) The denominators in each part of question 5 are called
   unrelated denominators. Why do you think they have this name?

c) When you add 2 fractions with unrelated denominators,
   how do you decide which number line to use?

7. Add.

a) $\frac{1}{3} + \frac{2}{7}$

b) $\frac{3}{4} + \frac{2}{9}$

c) $\frac{4}{5} + \frac{5}{8}$

d) $\frac{2}{5} + \frac{3}{7}$

8. Abey and Anoki are eating chocolate bars.
The bars are the same size.
Abey has $\frac{3}{4}$ left. Anoki has $\frac{5}{6}$ left.
How much chocolate is left altogether? Show your work.

9. **Assessment Focus** Use any of the digits
   1, 2, 3, 4, 5, 6 only once. Copy and complete.
   Replace each □ with a digit.

   $$\frac{\square}{\square} + \frac{\square}{\square}$$

   a) Find as many sums as you can that are between 1 and 2.
   b) Find the least sum that is greater than 1.
   Show your work.

10. Find 2 fractions with a sum of $\frac{3}{2}$. Try to do this as many ways as you can.
    Record each way you find.

11. **Take It Further** A jug holds 2 cups of liquid. A recipe for punch is
    $\frac{1}{2}$ cup of orange juice, $\frac{1}{4}$ cup of raspberry juice,
    $\frac{3}{8}$ cup of grapefruit juice, and $\frac{5}{8}$ cup of lemonade.
    Is the jug big enough for the punch? Explain how you know.

12. **Take It Further** A pitcher of juice is half empty.
    After $\frac{1}{2}$ cup of juice is added, the pitcher is $\frac{3}{4}$ full.
    How much juice does the pitcher hold when it is full?
    Show your thinking.

Music
Musical notes are named for fractions.
The type of note shows a musician how long to play
the note. In math, two halves make a whole — in
music, two half notes make a whole note!

| whole | half | quarter | eighth | sixteenth |
| note | note | note | note | note |

Reflect

What do you now know about adding fractions that you did
not know at the beginning of the lesson?

Focus   Use common denominators to
add fractions.

In Lessons 5.1 and 5.2, you used models to add fractions.
You may not always have suitable models.

You need a strategy you can use to add fractions
without using a model.

Explore

Copy these diagrams.

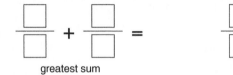

$$\frac{\square}{\square} + \frac{\square}{\square} =$$

greatest sum

$$\frac{\square}{\square} + \frac{\square}{\square} =$$

least sum

Use the digits 1, 2, 4, and 8 to make
the greatest sum and the least sum.
In each case, use each digit once.

Reflect & Share

Share your results with another pair of classmates.
Did you have the same answers?
If not, which is the greatest sum? The least sum?
What strategies did you use to add?

Connect

We can use equivalent fractions to add $\frac{1}{4} + \frac{1}{3}$.
Use equivalent fractions that have like denominators.
12 is a multiple of 3 and 4.
12 is a **common denominator**.
$$\frac{1}{4} = \frac{3}{12} \quad \text{and} \quad \frac{1}{3} = \frac{4}{12}$$
$$\text{So,} \frac{1}{4} + \frac{1}{3} = \frac{3}{12} + \frac{4}{12}$$
$$= \frac{7}{12}$$

Both fractions are written
with like denominators.

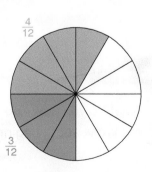

$\frac{4}{12}$

$\frac{3}{12}$

Look at the pattern in the equivalent fractions below.

$$\frac{1}{4} = \frac{3}{12} \qquad \frac{1}{3} = \frac{4}{12}$$

So, to get an equivalent fraction, multiply the numerator and denominator by the same number.

We may also get equivalent fractions by dividing.

For example, $\frac{8}{10}$ can be written: $\frac{8 \div 2}{10 \div 2} = \frac{4}{5}$

$\frac{8}{10}$ and $\frac{4}{5}$ are equivalent fractions.

$\frac{4}{5}$ is in simplest form.

Example

Add: $\frac{4}{9} + \frac{5}{6}$

Estimate to check the sum is reasonable.

A Solution

$\frac{4}{9} + \frac{5}{6}$

Estimate first.

$\frac{4}{9}$ is about $\frac{1}{2}$.

$\frac{5}{6}$ is close to 1.

So, $\frac{4}{9} + \frac{5}{6}$ is about $1\frac{1}{2}$.

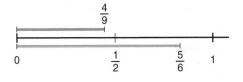

Use equivalent fractions to write the fractions with a common denominator.

List the multiples of 9: 9, **18**, 27, *36*, 45, …

List the multiples of 6: 6, 12, **18**, 24, 30, *36*, 42, …

18 is a multiple of 9 and 6, so 18 is a common denominator.

36 is also in both lists.
So, 36 is another possible
common denominator.

$$\frac{4}{9} = \frac{8}{18} \qquad \frac{5}{6} = \frac{15}{18}$$

$$\frac{4}{9} + \frac{5}{6} = \frac{8}{18} + \frac{15}{18} \qquad \text{Add the numerators.}$$
$$= \frac{23}{18}$$

We could have found this sum with fraction strips on a number line.

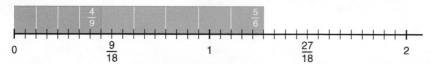

Since 23 > 18, this is an improper fraction.

To write the fraction as a mixed number:

$$\frac{23}{18} = \frac{18}{18} + \frac{5}{18}$$

$$= 1 + \frac{5}{18}$$

$$= 1\frac{5}{18} \qquad \text{This is a mixed number.}$$

The estimate was $1\frac{1}{2}$, so the answer is reasonable.

> **Recall that an improper fraction is a fraction with the numerator greater than the denominator.**

Practice

Write all sums in simplest form.

Write improper fractions as mixed numbers.

1. Find a common denominator for each pair of fractions.

   a) $\frac{1}{2}$ and $\frac{5}{8}$　　**b)** $\frac{1}{8}$ and $\frac{2}{3}$　　**c)** $\frac{2}{3}$ and $\frac{1}{9}$　　**d)** $\frac{3}{5}$ and $\frac{2}{3}$

2. Copy and complete. Replace each □ with a digit to make each equation true.

   a) $\frac{3}{12} = \frac{\square}{4}$　　**b)** $\frac{3}{4} = \frac{6}{\square}$　　**c)** $\frac{3}{6} = \frac{\square}{4}$　　**d)** $\frac{6}{8} = \frac{15}{\square}$

3. Add. Sketch a number line to model each sum.

   a) $\frac{4}{9} + \frac{1}{3}$　　**b)** $\frac{1}{2} + \frac{1}{3}$　　**c)** $\frac{3}{8} + \frac{3}{2}$　　**d)** $\frac{3}{4} + \frac{1}{6}$

4. Estimate, then add.

   a) $\frac{3}{5} + \frac{4}{8}$　　**b)** $\frac{1}{6} + \frac{5}{8}$　　**c)** $\frac{5}{6} + \frac{7}{9}$

   d) $\frac{3}{4} + \frac{4}{7}$　　**e)** $\frac{1}{3} + \frac{2}{5}$　　**f)** $\frac{1}{5} + \frac{5}{6}$

5. One page of a magazine had 2 advertisements.
   One was $\frac{1}{8}$ of the page, the other $\frac{1}{16}$.
   What fraction of the page was covered?
   Show your work.

6. Which sum is greater? Show your thinking.

$\frac{2}{3} + \frac{5}{6}$ or $\frac{3}{4} + \frac{4}{5}$

7. **Assessment Focus** Three people shared a pie.
Which statement is true? Can both statements be true?
Use pictures to show your thinking.

a) Edna ate $\frac{1}{10}$, Farrah ate $\frac{3}{5}$, and Ferris ate $\frac{1}{2}$.

b) Edna ate $\frac{3}{10}$, Farrah ate $\frac{1}{5}$, and Ferris ate $\frac{1}{2}$.

8. Damara and Baldwin had to shovel snow to clear their driveway.
Damara shovelled about $\frac{3}{10}$ of the driveway.
Baldwin shovelled about $\frac{2}{3}$ of the driveway.
What fraction of the driveway was cleared of snow?

9. Each fraction below is written as the sum of two unit fractions.
Which sums are correct? Why do you think so?

a) $\frac{7}{10} = \frac{1}{5} + \frac{1}{2}$ b) $\frac{5}{12} = \frac{1}{3} + \frac{1}{4}$ c) $\frac{5}{6} = \frac{1}{3} + \frac{1}{3}$

d) $\frac{7}{12} = \frac{1}{2} + \frac{1}{6}$ e) $\frac{11}{18} = \frac{1}{2} + \frac{1}{9}$ f) $\frac{2}{15} = \frac{1}{10} + \frac{1}{30}$

> A fraction with numerator 1 is a **unit fraction**.

10. **Take It Further** Add.

a) $\frac{3}{8} + \frac{1}{2} + \frac{3}{4}$ b) $\frac{1}{4} + \frac{3}{2} + \frac{2}{5}$ c) $\frac{2}{3} + \frac{5}{6} + \frac{4}{9}$

Reflect

Suppose your friend has forgotten how to add
two fractions with unlike denominators.
What would you do to help?

Mid-Unit Review

Write all sums in simplest form.
Write improper fractions as
mixed numbers.

5.1

1. Use fraction circles. Model this
picture, then find the sum.

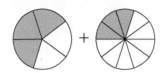

2. On Saturday, Howie hiked for
$\frac{5}{12}$ h in the morning and
$\frac{3}{6}$ h in the afternoon. What fraction
of an hour did Howie spend hiking?

5.2

3. Write an addition equation for
each picture.

a)

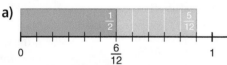

b)

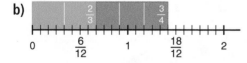

4. Add. Sketch fraction strips and a
number line to model each addition.

a) $\frac{2}{8} + \frac{3}{8}$ b) $\frac{2}{3} + \frac{1}{6}$

c) $\frac{3}{4} + \frac{2}{6}$ d) $\frac{1}{2} + \frac{2}{5}$

5. Find 3 different ways to add $\frac{2}{3} + \frac{5}{6}$.
Draw pictures to help you explain
each way.

5.3 **6.** Add. Estimate to check the sum
is reasonable.

a) $\frac{4}{8} + \frac{5}{8}$ b) $\frac{1}{3} + \frac{3}{5}$

c) $\frac{1}{4} + \frac{1}{8}$ d) $\frac{5}{6} + \frac{7}{12}$

7. Takoda and Wesley are collecting
shells on the beach in identical pails.
Takoda estimates she has filled
$\frac{7}{12}$ of her pail. Wesley estimates he
has filled $\frac{4}{10}$ of his pail. Suppose
the children combine their shells.
Will one pail be full? Explain.

8. Each guest at Tai's birthday party
brought one gift.
The circle graph shows the gifts
Tai received.

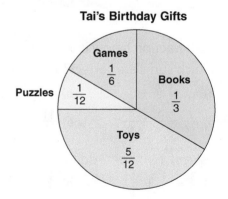

Tai's Birthday Gifts

a) What fraction of the gifts were:
   i)   toys or books?
   ii) puzzles or toys?
   iii) games or puzzles?
   iv) books or games?

b) Which 2 types of gifts represent
   $\frac{1}{4}$ of all the gifts?
   Explain how you know.

Using Models to Subtract Fractions

Focus Use Pattern Blocks, fraction strips, and number lines to subtract fractions.

Explore

You will need congruent squares, grid paper, and coloured pencils.

Use these rules to create a rectangular design.
The design must be symmetrical.
- One-half of the squares must be red.
- One-third of the squares must be blue.
- The remaining squares must be green.

What fraction of the squares are green? How do you know?
How many squares did you use?
Explain why you used that number of squares.
Describe your design.
Record your design on grid paper.

Reflect & Share

Compare your design with that of another pair of classmates.
If the designs are different, explain why your classmates' design obeys the rules.
How could you subtract fractions to find the fraction of the squares
that are green?

Connect

We can use models to subtract fractions.

To subtract $\frac{2}{3} - \frac{1}{2}$, we can use Pattern Blocks.
The yellow hexagon represents 1. The blue rhombus represents $\frac{1}{3}$.
The red trapezoid represents $\frac{1}{2}$.
Place 2 blue rhombuses over the hexagon.

To subtract $\frac{1}{2}$, place a red trapezoid over the 2 blue rhombuses.

Find a Pattern Block equal to the difference.
The green triangle represents the difference.

The green triangle is $\frac{1}{6}$ of the hexagon.
So, $\frac{2}{3} - \frac{1}{2} = \frac{1}{6}$

We can also use fraction strips and number lines to subtract.
To subtract fractions with unlike denominators, we use equivalent fractions.

Example
Subtract: $\frac{5}{8} - \frac{1}{4}$

A Solution
$\frac{5}{8} - \frac{1}{4}$
Think addition.
What do we add to $\frac{1}{4}$ to get $\frac{5}{8}$?
Use a number line that shows equivalent fractions
for eighths and fourths. That is, use the eighths number line.
Place the $\frac{1}{4}$-strip on the eighths number line with its right end at $\frac{5}{8}$.

Equivalent fractions:
$\frac{1}{4} = \frac{2}{8}$

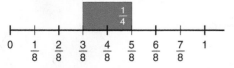

The left end of the strip is at $\frac{3}{8}$.
So, $\frac{5}{8} - \frac{1}{4} = \frac{3}{8}$

Use models.

1. Find equivalent fractions with like denominators for each pair of fractions.

a) $\frac{1}{2}$ and $\frac{5}{8}$

b) $\frac{1}{4}$ and $\frac{1}{3}$

c) $\frac{2}{3}$ and $\frac{1}{6}$

d) $\frac{3}{5}$ and $\frac{1}{2}$

2. Is each difference less than $\frac{1}{2}$ or greater than $\frac{1}{2}$? How can you tell?

a) $\frac{5}{6} - \frac{1}{2}$

b) $\frac{7}{8} - \frac{1}{8}$

c) $\frac{4}{6} - \frac{1}{3}$

d) $1 - \frac{5}{6}$

3. Subtract. Sketch pictures to show each difference.

a) $\frac{3}{4} - \frac{2}{4}$

b) $\frac{4}{5} - \frac{1}{5}$

c) $\frac{2}{3} - \frac{1}{3}$

d) $\frac{5}{8} - \frac{3}{8}$

4. a) Write a rule you could use to subtract fractions with like denominators without using number lines or fraction strips.

b) Write 3 subtraction questions with like denominators.
Use your rule to subtract the fractions.
Use fraction strips and number lines to check your answers.

5. Write a subtraction equation for each picture.

a)

b)

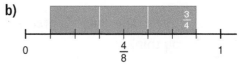

c)

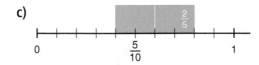

d)

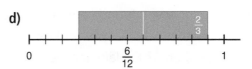

6. Subtract. Sketch pictures to show each difference.

a) $\frac{3}{8} - \frac{1}{4}$

b) $\frac{7}{10} - \frac{1}{2}$

c) $\frac{7}{8} - \frac{1}{2}$

d) $\frac{5}{6} - \frac{1}{4}$

7. Sergio has the lead role in the school play.
He still has to memorize $\frac{1}{2}$ of his lines.
Suppose Sergio memorizes $\frac{1}{3}$ of his lines today.
What fraction of his lines will he have left to memorize?
Show your work.

8. Freida has $\frac{3}{4}$ of a bottle of ginger ale.

She needs $\frac{1}{2}$ of a bottle of ginger ale for her fruit punch.

How much will be left in the bottle after Freida makes the punch?

9. A cookie recipe calls for $\frac{3}{4}$ cup of chocolate chips.

Spencer has $\frac{2}{3}$ cup. Does he have enough?

If your answer is yes, explain why it is enough.

If your answer is no, how much more does Spencer need?

10. Copy and replace each □ with a digit, to make each equation true.

Try to do this more than one way.

a) $\frac{2}{3} - \frac{\square}{\square} = \frac{1}{3}$ b) $\frac{\square}{\square} - \frac{1}{5} = \frac{3}{5}$ c) $\frac{\square}{3} - \frac{2}{\square} = \frac{1}{6}$

11. **Assessment Focus**   Kelly had $\frac{3}{4}$ of a tank of gas at

the beginning of the week.

At the end of the week, Kelly had $\frac{1}{8}$ of a tank left.

a) Did Kelly use more or less than $\frac{1}{2}$ of a tank? Explain.

b) How much more or less than $\frac{1}{2}$ of a tank did Kelly use?

Show your work.

12. a) Which of these differences is greater than $\frac{1}{2}$?

Why do you think so?

i) $\frac{5}{6} - \frac{2}{3}$ ii) $\frac{5}{6} - \frac{1}{2}$ iii) $\frac{5}{6} - \frac{1}{6}$

b) Explain how you found your answers to part a.

Which other way can you find the fractions with a difference

greater than $\frac{1}{2}$? Explain another strategy.

Reflect

When you subtract fractions with
unlike denominators, how do you subtract?
Give 2 different examples.
Use diagrams to show your thinking.

Focus | Use common denominators to subtract fractions.

Addition and subtraction are related operations.
You can use what you know about adding fractions to subtract them.

Explore

You will need fraction strips and number lines.
Find 2 fractions with a difference of $\frac{1}{2}$.
How many different pairs of fractions can you find?
Record each pair.

Reflect & Share

Discuss with your partner.
How are your strategies for subtracting fractions the same as your
strategies for adding fractions? How are they different?
Work together to use common denominators to subtract two fractions.

Connect

To subtract $\frac{4}{5} - \frac{1}{3}$, estimate first.
$\frac{4}{5}$ is close to 1, and $\frac{1}{3}$ is about $\frac{1}{2}$.
So, $\frac{4}{5} - \frac{1}{3}$ is about $1 - \frac{1}{2} = \frac{1}{2}$.
Use equivalent fractions to subtract.
Write $\frac{4}{5}$ and $\frac{1}{3}$ with a common denominator.

List the multiples of 5: 5, 10, **15**, 20, 25, …

List the multiples of 3: 3, 6, 9, 12, **15**, 18, …

15 is a multiple of 5 and 3, so 15 is a common denominator.

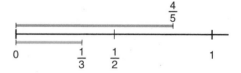

$$\frac{4}{5} \xrightarrow{\times 3} \frac{12}{15} \quad \text{and} \quad \frac{1}{3} \xrightarrow{\times 5} \frac{5}{15}$$

$\frac{4}{5} - \frac{1}{3} = \frac{12}{15} - \frac{5}{15}$

$\qquad = \frac{7}{15}$

**Think: 12 fifteenths minus
5 fifteenths is 7 fifteenths.**

We could have used a fraction strip on a number line.

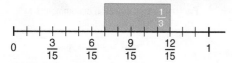

Example

Subtract.

a) $\frac{9}{10} - \frac{2}{5}$ b) $\frac{5}{4} - \frac{1}{5}$

Estimate to check the answer is reasonable.

A Solution

a) $\frac{9}{10} - \frac{2}{5}$

Estimate.

$\frac{9}{10}$ is about 1. $\frac{2}{5}$ is close to $\frac{1}{2}$.

So, $\frac{9}{10} - \frac{2}{5}$ is about $1 - \frac{1}{2} = \frac{1}{2}$.

Since 10 is a multiple of 5,
use 10 as a common denominator.

$$\frac{9}{10} - \frac{2}{5} = \frac{9}{10} - \frac{4}{10}$$

$$= \frac{5}{10} \qquad \text{This is not in simplest form.}$$

$$= \frac{5 \div 5}{10 \div 5} \qquad \text{5 is a factor of the numerator and denominator.}$$

$$= \frac{1}{2}$$

The estimate is $\frac{1}{2}$, so the difference is reasonable.

We could have used a fraction strip on a number line.

Another Strategy

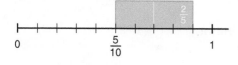

b) $\frac{5}{4} - \frac{1}{5}$

Estimate.

$\frac{5}{4}$ is about 1. $\frac{1}{5}$ is close to 0.

So, $\frac{5}{4} - \frac{1}{5}$ is about $1 - 0 = 1$.

Find a common denominator.

List the multiples of 4: 4, 8, 12, 16, **20**, 24, …

List the multiples of 5: 5, 10, 15, **20**, 25, …

20 is a multiple of 4 and 5, so 20 is a common denominator.

$$\overset{\times 5}{\frac{5}{4} = \frac{25}{20}} \underset{\times 5}{\qquad} \qquad\qquad \overset{\times 4}{\frac{1}{5} = \frac{4}{20}} \underset{\times 4}{\qquad}$$

$$\frac{5}{4} - \frac{1}{5} = \frac{25}{20} - \frac{4}{20}$$

$$= \frac{21}{20} \qquad\qquad \text{This is an improper fraction.}$$

$$\frac{21}{20} = \frac{20}{20} + \frac{1}{20}$$

$$= 1\frac{1}{20}$$

So, $\frac{5}{4} - \frac{1}{5} = 1\frac{1}{20}$

The estimate is 1, so the difference is reasonable.

Practice

Write all differences in simplest form.

1. Subtract.

a) $\frac{4}{5} - \frac{2}{5}$
b) $\frac{2}{3} - \frac{1}{3}$
c) $\frac{7}{9} - \frac{4}{9}$
d) $\frac{5}{7} - \frac{3}{7}$

2. Estimate, then subtract.

a) $\frac{2}{3} - \frac{1}{6}$
b) $\frac{5}{8} - \frac{1}{2}$
c) $\frac{3}{2} - \frac{7}{10}$
d) $\frac{11}{12} - \frac{5}{6}$

3. Subtract.

a) $\frac{3}{4} - \frac{2}{3}$
b) $\frac{4}{5} - \frac{2}{3}$
c) $\frac{7}{4} - \frac{4}{5}$
d) $\frac{3}{5} - \frac{1}{2}$

4. Subtract.

Estimate to check the answer is reasonable.

a) $\frac{4}{6} - \frac{1}{2}$
b) $\frac{5}{3} - \frac{3}{4}$
c) $\frac{7}{5} - \frac{5}{6}$
d) $\frac{5}{6} - \frac{3}{4}$

5. A recipe calls for $\frac{3}{4}$ cup of walnuts and $\frac{2}{3}$ cup of pecans.
Which type of nut is used more in the recipe?
How much more?

6. **Assessment Focus** On Saturday, Terri biked for $\frac{5}{6}$ h.

   On Sunday, Terri increased the time she biked by $\frac{7}{12}$ h.

   On Saturday, Bastien biked for $\frac{1}{2}$ h.

   On Sunday, Bastien increased the time he biked by $\frac{3}{4}$ h.

   a) Who biked longer on Sunday?
      How can you tell?

   b) For how much longer did this person bike?

   c) What did you need to know about fractions to
      answer these questions?

7. Write as many different subtraction questions
   as you can where the answer is $\frac{3}{4}$.
   Show your work.

8. The difference of 2 fractions is $\frac{1}{2}$.
   The lesser fraction is between 0 and $\frac{1}{4}$.
   What do you know about the other fraction?

9. **Take It Further** Meagan walks from home
   to school at a constant speed.
   It takes Meagan 3 min to walk the distance
   between Bonnie's house and Andrew's house.
   How long does it take Meagan to get to school?

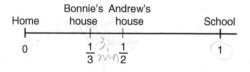

Reflect

Which fractions are easy to subtract?
Which are more difficult?
What makes them more difficult?
Give an example in each case.

5.6 Adding with Mixed Numbers

We have used fraction circles to model and add fractions.
We can also use fraction circles to model and add mixed numbers.
These fraction circles model $1\frac{5}{6}$.

1 whole $\frac{5}{6}$

Explore

Use any materials you want.
A recipe calls for $1\frac{1}{3}$ cups of all-purpose flour
and $\frac{5}{6}$ cup of whole-wheat flour.
How much flour is needed altogether?
How can you find out?
Show your work.

Reflect & Share

Describe your strategy.
Will your strategy work with all mixed numbers?
Test it with $2\frac{1}{3} + \frac{3}{4}$.
Use models or diagrams to justify your strategy.

Connect

Use fraction circles to add: $1\frac{3}{4} + 1\frac{3}{8}$

Use fraction circles to model $1\frac{3}{4}$ and $1\frac{3}{8}$.

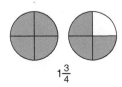
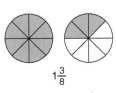

$1\frac{3}{4}$ $1\frac{3}{8}$

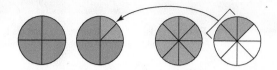

Use eighths to fill the fraction circle for $\frac{3}{4}$.

1 whole and 1 whole and 1 whole and 1 eighth equals 3 wholes and 1 eighth.

So, $1\frac{3}{4} + 1\frac{3}{8} = 3\frac{1}{8}$

To add with mixed numbers, we can:

- Add the fractions and add the whole numbers separately. Or:
- Write each mixed number as an improper fraction, then add.

Example

Add: $\frac{1}{3} + 1\frac{5}{6}$

A Solution

$\frac{1}{3} + 1\frac{5}{6}$

Estimate:

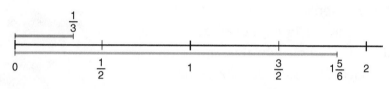

$1\frac{5}{6}$ is close to 2.

So, $\frac{1}{3} + 1\frac{5}{6} > 2$, but less than $2\frac{1}{3}$

Add the fractions and the whole number separately.

$\frac{1}{3} + 1\frac{5}{6} = \frac{1}{3} + \frac{5}{6} + 1$

Add the fractions: $\frac{1}{3} + \frac{5}{6}$

Since 6 is a multiple of 3, use 6 as a common denominator.

$$\frac{1}{3} \overset{\times 2}{\underset{\times 2}{=}} \frac{2}{6}$$

$\frac{1}{3} + \frac{5}{6} = \frac{2}{6} + \frac{5}{6}$

$\qquad = \frac{7}{6}$

Since 7 > 6, this is an improper fraction.

To write the improper fraction as a mixed number:

$\frac{7}{6} = \frac{6}{6} + \frac{1}{6}$

$\phantom{\frac{7}{6}} = 1 + \frac{1}{6}$

$\phantom{\frac{7}{6}} = 1\frac{1}{6}$

So, $\frac{1}{3} + \frac{5}{6} + 1 = 1\frac{1}{6} + 1$

$\phantom{So, \frac{1}{3} + \frac{5}{6} + 1} = 2\frac{1}{6}$

Then, $\frac{1}{3} + 1\frac{5}{6} = 2\frac{1}{6}$

This is close to the estimate of between 2 and $2\frac{1}{3}$,

so the sum is reasonable.

Another Solution

Write the mixed number as an improper fraction, then add.

$1\frac{5}{6} = 1 + \frac{5}{6}$

$\phantom{1\frac{5}{6}} = \frac{6}{6} + \frac{5}{6}$

$\phantom{1\frac{5}{6}} = \frac{11}{6}$

Since 6 is a multiple of 3, use 6 as a common denominator.

$$\frac{1}{3} \overset{\times 2}{\underset{\times 2}{=}} \frac{2}{6}$$

$\frac{1}{3} + 1\frac{5}{6} = \frac{2}{6} + \frac{11}{6}$

$\phantom{\frac{1}{3} + 1\frac{5}{6}} = \frac{13}{6}$

To write the fraction as a mixed number:

$\frac{13}{6} = \frac{12}{6} + \frac{1}{6}$

$\phantom{\frac{13}{6}} = 2 + \frac{1}{6}$

$\phantom{\frac{13}{6}} = 2\frac{1}{6}$

So, $\frac{1}{3} + 1\frac{5}{6} = 2\frac{1}{6}$

We can model this with a fraction strip on a number line.

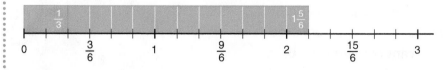

Write all sums in simplest form.

1. Write each mixed number as an improper fraction in simplest form.

a) $1\frac{3}{6}$ b) $4\frac{2}{8}$ c) $1\frac{3}{4}$ d) $3\frac{3}{5}$

2. Write each improper fraction as a mixed number in simplest form.

a) $\frac{17}{5}$ b) $\frac{9}{4}$ c) $\frac{18}{4}$ d) $\frac{28}{6}$

3. Use Pattern Blocks to find each sum.

a) $1\frac{1}{6} + \frac{2}{6}$ b) $1\frac{2}{3} + \frac{2}{3}$ c) $1\frac{4}{6} + 2\frac{1}{2}$ d) $2\frac{1}{3} + 3\frac{5}{6}$

4. Find each sum.

a) $3\frac{2}{3} + 2\frac{1}{3}$ b) $1\frac{1}{8} + 3\frac{5}{8}$ c) $4\frac{2}{9} + 3\frac{5}{9}$ d) $2\frac{3}{5} + 5\frac{4}{5}$

5. Use fraction circles to find each sum.

a) $2\frac{5}{8} + \frac{3}{4}$ b) $2\frac{5}{12} + \frac{2}{3}$ c) $1\frac{3}{8} + 3\frac{3}{4}$ d) $2\frac{2}{5} + 1\frac{7}{10}$

6. We know $\frac{1}{2} + \frac{1}{5} = \frac{7}{10}$.

Use this result to find each sum.

Estimate to check the sum is reasonable.

a) $3\frac{1}{2} + \frac{1}{5}$ b) $\frac{1}{2} + 2\frac{1}{5}$ c) $3\frac{1}{2} + 2\frac{1}{5}$ d) $4\frac{1}{2} + 3\frac{1}{5}$

7. For each pair of numbers, find a common denominator. Then add.

a) $3\frac{1}{3} + \frac{1}{4}$ b) $\frac{1}{2} + 1\frac{9}{10}$ c) $\frac{3}{4} + 2\frac{3}{5}$ d) $\frac{3}{7} + 2\frac{1}{2}$

e) $4\frac{7}{8} + 1\frac{2}{3}$ f) $2\frac{3}{5} + 2\frac{2}{3}$ g) $5\frac{2}{5} + 1\frac{7}{8}$ h) $3\frac{5}{6} + 2\frac{1}{4}$

8. Two students, Galen and Mai, worked on a project.

Galen worked for $3\frac{2}{3}$ h.

Mai worked for $2\frac{4}{5}$ h.

What was the total time spent on the project?

9. **Assessment Focus** Joseph used $1\frac{3}{8}$ cans of paint to paint

his room. Juntia used $2\frac{1}{4}$ cans to paint her room.

a) Estimate how many cans of paint were used in all.

b) Calculate how many cans of paint were used.

c) Draw a diagram to model your calculations in part b.

That is not a Dick

10. A recipe for punch calls for $2\frac{2}{3}$ cups of fruit concentrate and $6\frac{3}{4}$ cups of water.

How many cups of punch will the recipe make?

Show your work.

11. Use the fractions $1\frac{3}{5}$ and $2\frac{1}{10}$.

    a) Add the fractions and the whole numbers separately.

    b) Write each mixed number as an improper fraction.

    c) Add the improper fractions.

    d) Which method was easier: adding the mixed numbers or adding the improper fractions? Why do you think so? When would you use each method?

12. An auto mechanic completed 2 jobs before lunch.

The jobs took $2\frac{2}{3}$ h and $1\frac{3}{4}$ h.

How many hours did it take the mechanic to complete the 2 jobs?

13. **Take It Further** Replace the □ with an improper fraction or mixed number to make this equation true.

$3\frac{3}{5} + \square = 5$

Find as many answers as you can.

Draw diagrams to represent your thinking.

Reflect

How is adding a mixed number and a fraction like adding two fractions?

How is it different?

Use examples to explain.

Subtracting with Mixed Numbers

We can use Cuisenaire rods to model fractions and mixed numbers.
Suppose the dark green rod is 1 whole, then the red rod is $\frac{1}{3}$.
So, seven red rods is $\frac{7}{3}$, or $2\frac{1}{3}$.

$\frac{1}{3}$	$\frac{1}{3}$	$\frac{1}{3}$

1

1		
$\frac{1}{3}$	$\frac{1}{3}$	$\frac{1}{3}$
$\frac{1}{3}$	$\frac{1}{3}$	$\frac{1}{3}$
$\frac{1}{3}$		

$2\frac{1}{3}$

Explore

Use any materials you want.
A bicycle shop closed for lunch for $1\frac{2}{3}$ h on Monday and for $\frac{3}{4}$ h on Tuesday.
How much longer was the shop closed for lunch on Monday than on Tuesday?
How can you find out? Show your work.

Reflect & Share

Describe your strategy.
Will your strategy work with all mixed numbers?
Test it with $2\frac{1}{4} - \frac{3}{8}$.
Use models or diagrams to justify your strategy.

Connect

Use Cuisenaire rods to subtract: $1\frac{1}{2} - \frac{3}{4}$

Use Cuisenaire rods to model $1\frac{1}{2}$ and $\frac{3}{4}$.

Let the brown rod represent 1 whole.

Then, the purple rod represents $\frac{1}{2}$ and the red rod represents $\frac{1}{4}$.

Model $1\frac{1}{2}$ with Cuisenaire rods.

1	$\frac{1}{2}$

Model $\frac{3}{4}$ with Cuisenaire rods.

Place the rods for $\frac{3}{4}$ above the rods for $1\frac{1}{2}$, so they align at the right.

Find a rod equal to the difference in their lengths.
The difference is equal to the dark green rod.

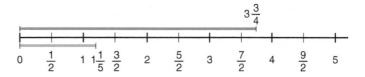

The dark green rod represents $\frac{3}{4}$ of the brown rod.
So, $1\frac{1}{2} - \frac{3}{4} = \frac{3}{4}$
To subtract with mixed numbers, we can:
- Subtract the fractions and subtract the whole numbers separately. Or:
- Write each mixed number as an improper fraction, then subtract.

Example

Subtract.

a) $3\frac{3}{4} - 1\frac{1}{5}$ **b)** $3\frac{1}{5} - \frac{3}{4}$

Estimate to check the answer is reasonable.

A Solution

a) $3\frac{3}{4} - 1\frac{1}{5}$

Estimate.

$3\frac{3}{4}$ is about 4. $1\frac{1}{5}$ is about 1.

So, $3\frac{3}{4} - 1\frac{1}{5}$ is between 2 and 3.

Subtract the fractions first: $\frac{3}{4} - \frac{1}{5}$

The denominators 4 and 5
have no common factors.
So, a common denominator is: $4 \times 5 = 20$.

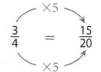

$$\frac{3}{4} - \frac{1}{5} = \frac{15}{20} - \frac{4}{20}$$
$$= \frac{11}{20}$$

Subtract the whole numbers: $3 - 1 = 2$

Then, $3\frac{3}{4} - 1\frac{1}{5} = 2\frac{11}{20}$

This is close to the estimate of between 2 and 3,

so the answer is reasonable.

b) $3\frac{1}{5} - \frac{3}{4}$

Estimate.

$3\frac{1}{5}$ is about 3.

$\frac{3}{4}$ is close to 1.

So, $3\frac{1}{5} - \frac{3}{4}$ is about $3 - 1 = 2$.

We cannot subtract the fractions because $\frac{1}{5} < \frac{3}{4}$.

So, write $3\frac{1}{5}$ as an improper fraction.

$$3\frac{1}{5} = 3 + \frac{1}{5}$$
$$= \frac{15}{5} + \frac{1}{5}$$
$$= \frac{16}{5}$$

> **Another Strategy**
> We could use fraction
> circles to subtract.

The denominators have no common factors.

So, a common denominator is: $4 \times 5 = 20$

$$\frac{16}{5} \overset{\times 4}{\underset{\times 4}{=}} \frac{64}{20} \qquad \frac{3}{4} \overset{\times 5}{\underset{\times 5}{=}} \frac{15}{20}$$

$$\frac{16}{5} - \frac{3}{4} = \frac{64}{20} - \frac{15}{20}$$
$$= \frac{49}{20}$$
$$= \frac{40}{20} + \frac{9}{20}$$
$$= 2 + \frac{9}{20}$$
$$= 2\frac{9}{20}$$

So, $3\frac{1}{5} - \frac{3}{4} = 2\frac{9}{20}$

This is close to the estimate of 2, so the answer is reasonable.

Before we subtract the fraction parts of two mixed numbers, we must check the fractions to see which is greater. When the second fraction is greater than the first fraction, we cannot subtract directly.

Practice

Write all differences in simplest form.

1. Subtract.

a) $2\frac{3}{5} - 1\frac{2}{5}$ b) $3\frac{7}{8} - 1\frac{5}{8}$ c) $\frac{15}{4} - \frac{3}{4}$ d) $\frac{11}{6} - \frac{1}{6}$

2. Subtract. Use Cuisenaire rods.
Sketch diagrams to record your work.

a) $1\frac{2}{3} - \frac{2}{6}$ b) $3\frac{1}{2} - 1\frac{2}{4}$ c) $3\frac{3}{10} - 2\frac{4}{5}$ d) $2\frac{1}{4} - \frac{1}{2}$

3. We know that $\frac{2}{3} - \frac{1}{2} = \frac{1}{6}$.
Use this result to find each difference.
Estimate to check the answer is reasonable.

a) $2\frac{2}{3} - \frac{1}{2}$ b) $2\frac{2}{3} - 1\frac{1}{2}$ c) $4\frac{2}{3} - 2\frac{1}{2}$ d) $5\frac{2}{3} - 1\frac{1}{2}$

4. Estimate, then subtract.

a) $\frac{7}{2} - \frac{5}{4}$ b) $\frac{13}{6} - \frac{8}{12}$ c) $\frac{5}{4} - \frac{3}{5}$ d) $\frac{9}{5} - \frac{1}{2}$

5. a) Subtract.

i) $3 - \frac{4}{5}$ ii) $4 - \frac{3}{7}$ iii) $5 - \frac{5}{6}$ iv) $6 - \frac{4}{9}$

b) Which methods did you use in part a?
Explain your choice.

6. For the fractions in each pair of numbers, find a common denominator.
Then subtract.

a) $3\frac{3}{4} - 1\frac{1}{5}$ b) $4\frac{9}{10} - 3\frac{1}{2}$ c) $3\frac{3}{4} - 1\frac{1}{3}$ d) $4\frac{5}{7} - 2\frac{2}{3}$

7. For each pair of mixed numbers below:

a) Subtract the fractions and subtract the whole numbers separately.

b) Write the mixed numbers as improper fractions, then subtract.

c) Which method was easier? Why do you think so?

i) $3\frac{3}{5} - 1\frac{3}{10}$ ii) $3\frac{3}{10} - 1\frac{3}{5}$

8. A flask contains $2\frac{1}{2}$ cups of juice.

Ping drinks $\frac{3}{8}$ cup of juice, then Preston drinks $\frac{7}{10}$ cup of juice.

How much juice is in the flask now? Show your work.

9. The running time of a movie is $2\frac{1}{6}$ h.

In the theatre, Jason looks at his watch and sees that $1\frac{1}{4}$ h has passed.

How much longer will the movie run?

10. Subtract.

a) $3\frac{2}{3} - 2\frac{7}{8}$ b) $5\frac{1}{2} - 3\frac{7}{9}$ c) $4\frac{3}{5} - 1\frac{2}{3}$ d) $4\frac{2}{5} - 1\frac{7}{8}$

11. Assessment Focus The students in two Grade 7 classes made sandwiches for parents' night.

Mr. Crowe's class used $5\frac{1}{8}$ loaves of bread.

Mme. Boudreau's class used $3\frac{2}{3}$ loaves of bread.

a) Estimate how many more loaves Mr. Crowe's class used.

b) Calculate how many more loaves Mr. Crowe's class used.

c) Draw a diagram to model your calculations in part b.

d) The two classes purchased 10 loaves.

How many loaves were left?

12. Take It Further Replace the □ with an improper fraction or mixed number to make this equation true.

$4\frac{1}{8} - \square = 1\frac{1}{2}$

Find as many answers as you can.

Draw diagrams to represent your thinking.

Reflect

You have learned to use improper fractions to subtract mixed numbers.
When is this not the better method? Use an example to explain.

Magazines and newspapers make money by selling advertising space.

The advertising sales representative contacts companies whose products might be of interest to readers. She offers to sell them various sizes of advertisement space at different rates.
When talking about ads smaller than a full page, the sales rep uses fractions to describe them. It's much simpler to talk about a $\frac{2}{3}$-page ad instead of a 0.666 667 page ad!

The sales rep tries to sell combinations of ads that can fill pages, with no space left over. A sales rep has sold two $\frac{1}{4}$-page ads and one $\frac{1}{6}$-page ad. She wants to know the possible combinations of ad sizes she can sell to fill the rest of the page.
What might they be?

Writing a Complete Solution

A question often says "Show your work."
What does this mean?

When you are asked to show your work,
you should show your thinking by writing
a complete solution.

Work with a partner.
Compare these solutions.

Solution 1

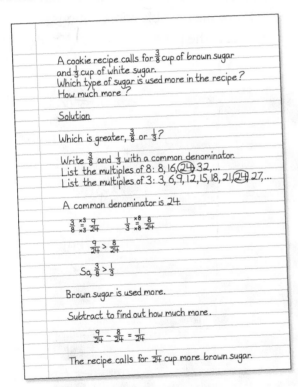

A cookie recipe calls for $\frac{3}{8}$ cup of brown sugar
and $\frac{1}{3}$ cup of white sugar.
Which type of sugar is used more in the recipe?
How much more?

Solution

Which is greater, $\frac{3}{8}$ or $\frac{1}{3}$?

Write $\frac{3}{8}$ and $\frac{1}{3}$ with a common denominator.
List the multiples of 8: 8, 16, 24, 32,...
List the multiples of 3: 3, 6, 9, 12, 15, 18, 21, 24, 27,...

A common denominator is 24.

$\frac{3}{8} = \frac{\times 3}{\times 3} = \frac{9}{24}$ $\frac{1}{3} = \frac{\times 8}{\times 8} = \frac{8}{24}$

$\frac{9}{24} > \frac{8}{24}$

So, $\frac{3}{8} > \frac{1}{3}$

Brown sugar is used more.

Subtract to find out how much more.

$\frac{9}{24} - \frac{8}{24} = \frac{1}{24}$

The recipe calls for $\frac{1}{24}$ cup more brown sugar.

Solution 2

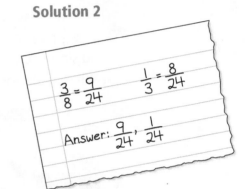

$\frac{3}{8} = \frac{9}{24}$ $\frac{1}{3} = \frac{8}{24}$

Answer: $\frac{9}{24}$, $\frac{1}{24}$

- Which solution is complete?
- Suppose this question is on a test. It is worth 4 marks.
  How many marks would you give each solution above?
  Justify your answers.

Make a list of things that should be included in a complete solution.

Tips for writing a complete solution:

- Write down the question.
- Show all steps so that someone else can follow your thinking.
- Include graphs or pictures to help explain your thinking.
- Check that your calculations are accurate.
- Use math symbols correctly.
- Write a concluding sentence that answers the question.

Name: _____   Date: _____

1. Which fraction is greater? How do you know?

   $\frac{3}{5}$, $\frac{2}{3}$

2. Add.

   $\frac{3}{4} + \frac{1}{12}$

3. Subtract.

   $4\frac{5}{6} - \frac{17}{18}$

4. Marty drank $\frac{4}{5}$ cup of orange juice.

   Kobe drank $\frac{3}{4}$ cup of orange juice.

   a) Who drank more orange juice?
   b) How much more orange juice did he drink?

Unit Review

Adding and Subtracting Fractions

✓ Use models, such as Pattern Blocks, fraction circles, fraction strips, and number lines.

✓ Like denominators: add or subtract the numerators.

For example, $\frac{5}{6} + \frac{2}{6} = \frac{7}{6}$ $\frac{5}{6} - \frac{2}{6} = \frac{3}{6}$, or $\frac{1}{2}$

✓ Unlike denominators: Use a common denominator to write equivalent fractions, then add or subtract the numerators.

For example:

$$\frac{3}{4} + \frac{3}{5}$$
$$= \frac{15}{20} + \frac{12}{20}$$
$$= \frac{27}{20}, \text{ or } 1\frac{7}{20}$$

$$\frac{3}{4} - \frac{3}{5}$$
$$= \frac{15}{20} - \frac{12}{20}$$
$$= \frac{3}{20}$$

Adding and Subtracting with Mixed Numbers

✓ Use models, such as fraction circles, Pattern Blocks, and Cuisenaire rods.

✓ Add or subtract the fractions and the whole numbers separately.

For example:

$$3\frac{5}{8} + 2\frac{1}{4}$$
$$= 3 + 2 + \frac{5}{8} + \frac{1}{4}$$
$$= 5 + \frac{5}{8} + \frac{2}{8}$$
$$= 5 + \frac{7}{8}$$
$$= 5\frac{7}{8}$$

$$3\frac{2}{3} - 1\frac{3}{5}$$
$$= 3 - 1 + \frac{2}{3} - \frac{3}{5}$$
$$= 2 + \frac{10}{15} - \frac{9}{15}$$
$$= 2 + \frac{1}{15}$$
$$= 2\frac{1}{15}$$

Check that $\frac{2}{3} > \frac{3}{5}$.

✓ Write each mixed number as an improper fraction, then add or subtract.

For example:

$$1\frac{5}{6} + 1\frac{2}{5}$$
$$= \frac{11}{6} + \frac{7}{5}$$
$$= \frac{55}{30} + \frac{42}{30}$$
$$= \frac{97}{30}, \text{ or } 3\frac{7}{30}$$

$$2\frac{1}{4} - 1\frac{1}{2}$$
$$= \frac{9}{4} - \frac{3}{2}$$
$$= \frac{9}{4} - \frac{6}{4}$$
$$= \frac{3}{4}$$

Since $\frac{1}{4} < \frac{1}{2}$, use improper fractions.

What Should I Be Able to Do?

LESSON

5.1 **1.** Add.

Use fraction circles.

Draw a picture to show each sum.

a) $\frac{8}{12} + \frac{5}{12}$ b) $\frac{3}{4} + \frac{2}{8}$

c) $\frac{1}{4} + \frac{2}{3}$ d) $\frac{1}{10} + \frac{3}{5}$

5.2 **2.** Add. Use fraction strips on number lines.

Draw a picture to show each sum.

a) $\frac{5}{9} + \frac{2}{3}$ b) $\frac{2}{3} + \frac{5}{6}$

c) $\frac{1}{6} + \frac{7}{12}$ d) $\frac{3}{8} + \frac{6}{8}$

3. Find 2 fractions that add to $\frac{5}{8}$.
Find as many pairs of fractions as you can.

5.3 **4.** Find a common denominator for each set of fractions.
Write equivalent fractions for each pair.

a) $\frac{3}{5}$ and $\frac{3}{4}$ b) $\frac{2}{5}$ and $\frac{3}{15}$

c) $\frac{4}{9}$ and $\frac{1}{2}$ d) $\frac{5}{8}$ and $\frac{1}{6}$

5. Add.

a) $\frac{1}{5} + \frac{3}{5}$ b) $\frac{1}{2} + \frac{3}{7}$

c) $\frac{2}{3} + \frac{3}{10}$ d) $\frac{3}{5} + \frac{1}{4}$

5.4 **6.** Write a subtraction equation for each picture.

a)

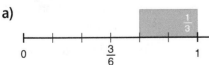

b)

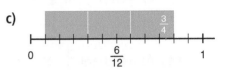

c)

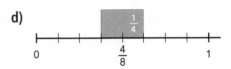

d)

7. Subtract. Draw a picture to show each difference.

a) $\frac{4}{5} - \frac{1}{5}$

b) $\frac{5}{6} - \frac{1}{3}$

c) $\frac{11}{12} - \frac{1}{2}$

8. Joyce and Javon each have the same MP3 player.
Joyce has used $\frac{7}{9}$ of her storage capacity.
Javon has used $\frac{5}{6}$ of his storage capacity.

a) Who has used more storage capacity?

b) How much more storage capacity has he or she used?

Show your work.

5.5 **9.** Subtract.

a) $\frac{9}{10} - \frac{2}{5}$ b) $\frac{7}{3} - \frac{5}{6}$

c) $\frac{8}{5} - \frac{1}{4}$ d) $\frac{9}{4} - \frac{2}{3}$

10. Write a subtraction question that has each fraction below as the answer.
The two fractions that are subtracted should have unlike denominators.

a) $\frac{1}{2}$ b) $\frac{3}{4}$ c) $\frac{1}{10}$

d) $\frac{1}{6}$ e) $\frac{1}{4}$

11. Anton drank $\frac{3}{4}$ bottle of water.
Brad drank $\frac{7}{8}$ bottle of water.

a) Who drank more water?

b) How much more water did he drink?

12. The gas tank in Eddie's car is $\frac{5}{8}$ full. He uses $\frac{1}{4}$ tank of gas to run his errands. What fraction of a tank of gas is left?

5.6 **13.** Use fraction circles to find each sum.

a) $6\frac{1}{3} + \frac{1}{3}$ b) $1\frac{5}{12} + \frac{1}{6}$

c) $2\frac{3}{10} + 3\frac{1}{5}$ d) $5\frac{1}{4} + 1\frac{2}{5}$

14. Add.

a) $3\frac{5}{6} + \frac{4}{6}$ b) $4\frac{3}{8} + \frac{1}{4}$

c) $7\frac{3}{10} + 2\frac{4}{5}$ d) $2\frac{5}{9} + 5\frac{2}{3}$

15. Danielle mows lawns as a part-time job. On Monday, Danielle spent $1\frac{3}{4}$ h mowing lawns.
On Wednesday, she spent $1\frac{7}{8}$ h mowing lawns.
How much time did she spend mowing lawns over the 2 days?

5.7 **16.** Subtract. Draw a picture to show each difference.

a) $4\frac{1}{2} - \frac{3}{8}$ b) $3\frac{4}{9} - \frac{2}{3}$

c) $5\frac{5}{12} - 3\frac{5}{6}$ d) $4\frac{5}{8} - 2\frac{2}{3}$

17. Amelie wants to bake two kinds of muffins. One recipe calls for $1\frac{3}{4}$ cups of bananas. The other recipe calls for $1\frac{7}{8}$ cups of cranberries.

a) Which recipe uses more fruit?

b) How much more fruit does the recipe in part a use?

18. Add or subtract as indicated.

a) $2\frac{2}{3} + 1\frac{1}{2}$ b) $3\frac{1}{3} - 1\frac{7}{10}$

c) $2\frac{1}{6} + 4\frac{7}{8}$ d) $3\frac{1}{2} - 2\frac{3}{4}$

19. On a trip from Edmonton to Saskatoon, Carly drove for $2\frac{1}{2}$ h, stopped for gas and lunch, then drove for $2\frac{2}{3}$ h.
The total trip took 6 h. How long did Carly stop for gas and lunch? Express your answer as a fraction of an hour.

Practice Test

1. Add or subtract.
   Draw a picture to show each sum or difference.
   Write each sum or difference in simplest form.

   a) $\frac{7}{5} + \frac{3}{5}$ b) $\frac{13}{10} - \frac{2}{3}$ c) $\frac{11}{12} - \frac{8}{12}$ d) $\frac{4}{9} + \frac{7}{6}$

2. Find two fractions that have a sum of $\frac{3}{5}$.
   a) The fractions have like denominators.
   b) The fractions have unlike denominators.

3. Find two fractions that have a difference of $\frac{1}{4}$.
   a) The fractions have like denominators.
   b) The fractions have unlike denominators.

4. Add or subtract.

   a) $6\frac{3}{8} + 2\frac{1}{5}$ b) $3\frac{1}{10} - 1\frac{4}{5}$

5. Lana does yard work.
   The table shows the approximate time for each job.
   For one Saturday, Lana has these jobs:
   – mow 3 small lawns
   – mow 1 large lawn
   – mow lawn/tidy yard in 2 places
   – plant annuals in 1 place
   Lana needs travel time between jobs,
   and a break for lunch.
   Do you think she will be able to do all the jobs? Justify your answer.

Job	Time
Mow small lawn	$\frac{1}{2}$ h
Mow large lawn	$\frac{3}{4}$ h
Mow lawn/tidy yard	$1\frac{1}{2}$ h
Plant annuals	$2\frac{1}{2}$ h

6. Write each fraction as the sum of two different unit fractions.

   a) $\frac{3}{4}$ b) $\frac{5}{8}$

7. A fraction is written on each side of two counters.
   All the fractions are different.
   The counters are flipped and the fractions are added.
   Their possible sums are: $1, 1\frac{1}{4}, \frac{7}{12}, \frac{5}{6}$
   Which fractions are written on the counters?
   Explain how you found the fractions.

The students at Anishwabe School are preparing a
special book for the school's 100th anniversary.
They finance the book by selling advertising space to sponsors.
The students sold the following space:

Full page	$\frac{1}{2}$ page	$\frac{1}{3}$ page	$\frac{1}{4}$ page	$\frac{1}{6}$ page	$\frac{1}{8}$ page
1	1	1	3	4	5

All the advertisements are to fit at the back of the book.
Sam asks: "How many pages do we need for the advertisements?"
Ruth asks: "Will the advertisements fill the pages?"
Jiba asks: "Is there more than one way to arrange
these advertisements?"
Can you think of other questions students might ask?

1. Find the total advertising space needed.
2. Sketch the pages to find how the advertisements can be placed.
   Use grid paper if it helps.

3. Compare your group's sketch with those of other groups.
When you made your sketch, what decisions did you make
about the shape of each advertisement?
Did other groups make the same decisions?
If your answer is no, explain how another group made its decisions.

4. What are the fewest pages needed to display the advertisements?
Will there be room for any other advertisements?
How can you tell?

Check List

Your work
should show:

✓ all calculations
in detail

✓ diagrams of the
layout for the
advertisements

✓ a clear explanation
of how you prepared
the layout

✓ a clear explanation
of which students
received the prizes,
and how much more
the third student
needed to sell

5. What else might students need to consider as they prepare
the layout for the book?

To encourage students to sell advertisements, the organizing
committee offered prizes to the 2 students who sold the most space.

Sandra, Roy, and Edward are the top sellers.
This table shows the advertising space each of these students sold.

	Full page	$\frac{1}{2}$ page	$\frac{1}{3}$ page	$\frac{1}{4}$ page	$\frac{1}{6}$ page	$\frac{1}{8}$ page
Sandra		1			1	2
Roy	1				1	
Edward			1	1	1	1

6. Which two students sold the most space?
Show how you know.

7. How much more space would the third-place student
have to sell to receive first prize?
Second prize? Show your work.

Reflect on Your Learning

Look back at the goals under *What You'll Learn*.
How well do you think you have met these goals?

Equations

Many digital music clubs offer albums for subscribers to download.

These tables show the plans for downloading albums for two companies. Each plan includes 5 free album downloads per month.

What patterns do you see in the tables? Write a pattern rule for each pattern. Describe each plan.

Which is the more expensive plan for 8 additional albums? Assume the patterns continue. Will this company always be more expensive? How do you know?

What You'll Learn

- Demonstrate and use the preservation of equality.
- Explain the difference between an expression and an equation.
- Use models, pictures, and symbols to solve equations and verify the solutions.
- Solve equations using algebra.
- Make decisions about which method to use to solve an equation.
- Solve problems using related equations.

Why It's Important

- Using equations is an effective problem-solving tool.
- Using algebra to solve equations plays an important role in many careers. For example, urban planners use equations to investigate population growth.

systematic trial
inspection

Company A

Number of Additional Albums Downloaded per Month	Total Cost ($)
0	10
1	18
2	26
3	34
4	42
5	50

Company B

Number of Additional Albums Downloaded per Month	Total Cost ($)
0	20
1	27
2	34
3	41
4	48
5	55

Look at the algebraic expressions and equations below.
Which ones are equations? Which ones are expressions?
How do you know?

$3n + 12$ $3n = 12$ $5x + 2$ $5x + 2 = 27$

Explore

On the way home from school,
10 students got off the bus at the first stop.
There were then 16 students on the bus.
How many students were on the bus
when it left the school?
How many different ways can you solve the problem?

Reflect & Share

Discuss your strategies for finding the answer
with another pair of classmates.
Did you use an equation?
Did you use reasoning?
Did you draw a picture?
Justify your choice.

Connect

Janet walked a total of 17 km in February.
She walked the same number of kilometres
in each of the first 3 weeks.
Then she walked 5 km in the fourth week.
How many kilometres did Janet walk
in each of the first 3 weeks?

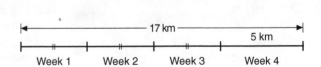

Let d represent the distance Janet walked, in kilometres, in each of the first 3 weeks.
So $3 \times d$, or $3d$, represents the total number of kilometres Janet walked in the first 3 weeks.
She walked 5 km in the fourth week, for a total of 17 km.
The equation is: $3d + 5 = 17$

When we use the equation to find the value of d, we *solve the equation*.

Here are 2 ways to solve this equation.

Writing and solving equations is a useful strategy for solving problems.

Method 1: By Systematic Trial

$3d + 5 = 17$
We choose a value for d and substitute.

Try $d = 2$.
$$\begin{aligned} 3d + 5 &= 3 \times 2 + 5 \\ &= 6 + 5 \\ &= 11 \end{aligned}$$

11 is too small, so choose a greater value for d.

Try $d = 5$.
$$\begin{aligned} 3d + 5 &= 3 \times 5 + 5 \\ &= 15 + 5 \\ &= 20 \end{aligned}$$

20 is too large, so choose a lesser value for d.

Try $d = 4$.
$$\begin{aligned} 3d + 5 &= 3 \times 4 + 5 \\ &= 12 + 5 \\ &= 17 \end{aligned}$$

This is correct.
Janet walked 4 km during each of the first 3 weeks of February.

Systematic trial means choosing a value for the variable, then checking by substituting. Use the answer and reasoning to choose the next value to check.

Method 2: By Inspection

$3d + 5 = 17$
We first find a number which, when added to 5, gives 17.
$$3d + 5 = 17$$
We know that $12 + 5 = 17$.
So, $3d = 12$
Then we find a number which, when multiplied by 3, has product 12.
We know that $3 \times 4 = 12$; so $d = 4$
Janet walked 4 km during each of the first 3 weeks of February.

By inspection means finding the value for the variable by using these types of number facts: addition, subtraction, multiplication, and division

We say that the value $d = 4$ makes the equation $3d + 5 = 17$ true.
Any other value of d, such as $d = 6$,
would not make the equation true.

$3 \times 6 + 5$ does not equal 17.

The value $d = 4$ is the only solution to the equation.
That is, there is only one value of d that makes the equation true.

Example

For each situation, write an equation.
Ben has a large collection of baseball caps.

a) Ben takes y caps from a group of 18 caps.
   There are 12 caps left.
   How many caps did Ben take away?
   Solve the equation by inspection.

b) Ben put k caps in each of 6 piles.
   There are 108 caps altogether.
   How many caps did Ben put in each pile?
   Solve the equation by systematic trial.

c) Ben shares n caps equally among 9 piles.
   There are 6 caps in each pile.
   How many caps did Ben have?
   Solve the equation by inspection.

d) Ben combines p groups of 4 caps each into one large group.
   He then takes away 7 caps. There are 49 caps left.
   How many groups of 4 caps did Ben begin with?
   Solve the equation by systematic trial.

A Solution

a) 18 subtract y equals 12.
   $$18 - y = 12$$
   Which number subtracted from 18 gives 12?
   We know that $18 - 6 = 12$; so $y = 6$.
   Ben took away 6 caps.

b) 6 times k equals 108.
   $$6k = 108$$
   Try $k = 15$. $6k = 6(15)$

   Recall: $6(15) = 6 \times 15$

                      $= 90$
   90 is too small, so choose a greater value for k.

Try $k = 20$.　　　$6k = 6(20)$
　　　　　　　　　$= 120$

120 is too large, so choose a lesser value for k.

Try $k = 17$.　　　$6k = 6(17)$
　　　　　　　　　$= 102$

102 is too small, but it is close to the value we want.

Try $k = 18$.　　　$6k = 6(18)$
　　　　　　　　　$= 108$

This is correct.

Ben put 18 caps in each pile.

c) n divided by 9 equals 6.

$n \div 9 = 6$, or $\frac{n}{9} = 6$

Which number divided by 9 gives 6?

We know that $54 \div 9 = 6$; so $n = 54$.

Ben had 54 caps.

d) 4 times p subtract 7 equals 49.

$4p - 7 = 49$

Since $4 \times 10 = 40$, we know we need to start with a value for p greater than 10.

Try $p = 12$.　　　$4p - 7 = 4 \times 12 - 7$
　　　　　　　　　　$= 48 - 7$
　　　　　　　　　　$= 41$, which is too small

41 is 8 less than 49, so we need two more groups of 4.

Try $p = 14$.　　　$4p - 7 = 4 \times 14 - 7$
　　　　　　　　　　$= 56 - 7$
　　　　　　　　　　$= 49$

This is correct.

Ben began with 14 groups of 4 caps each.

Practice

1. Look at the algebraic expressions and equations below.
Which are expressions? Equations?
How do you know?

a) $4w = 48$　　　　　b) $g - 11$　　　　　c) $3d + 5$

d) $\frac{x}{12} = 8$　　　　e) $\frac{j - 5}{10}$　　　　f) $6z + 1 = 67$

2. Solve each equation in question 1 by inspection or by systematic trial. Explain why you chose the method you did.

3. Shenker gives 10 CDs to his brother.
Shenker then has 35 CDs.
   a) Write an equation you can solve to find how many CDs Shenker had to begin with.
   b) Solve the equation.

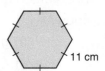

4. Write an equation for each sentence.
Solve each equation by inspection.
   a) Seven more than a number is 18.
   b) Six less than a number is 24.
   c) Five times a number is 45.
   d) A number divided by six is 7.
   e) Three more than four times a number is 19.

5. Write an equation you could use to solve each problem.
Solve each equation by systematic trial.
   a) Aiko bought 14 DVDs for $182.
      She paid the same amount for each DVD.
      How much did each DVD cost?
   b) Kihew collects beaded leather bracelets. She lost 14 of her bracelets.
      Kihew has 53 bracelets left.
      How many bracelets did she have to begin with?
   c) Manuel gets prize points for reading books.
      He needs 100 points to win a set of tangrams.
      Manuel has 56 points. When he reads 11 more books,
      he will have 100 points.
      How many points does Manuel get for each book he reads?

6. The perimeter of a square is 48 cm.
   a) Write an equation you can solve to find the side length of the square.
   b) Solve the equation.

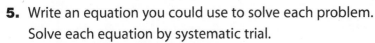

7. The side length of a regular hexagon is 11 cm.
   a) Write an equation you can solve to find the perimeter of the hexagon.
   b) Solve the equation.

11 cm

8. Use questions 6 and 7 as a guide.

  a) Write your own problem about side length and perimeter of a figure.

  b) Write an equation you can use to solve the problem.

  c) Solve the equation.

9.  Eli has 130 key chains. He keeps 10 key chains for himself, then shares the rest equally among his friends.
Each friend then has 24 key chains.

  a) Write an equation you can solve to find
  how many friends were given key chains.

  b) Solve the equation by inspection, then by systematic trial.
  Which method was easier to use? Explain your choice.

10. Find the value of n that makes each equation true.

  a) $3n = 27$ **b)** $2n + 3 = 27$ **c)** $2n - 3 = 27$ **d)** $\frac{n}{3} = 27$

11. **Take It Further** Write a problem that can be described by each equation.
Solve each equation. Which equation was the most difficult to solve?
Why do you think so?

  a) $2x - 1 = 5$ **b)** $4y = 24$

  c) $\frac{z}{38} = 57$ **d)** $5x + 5 = 30$

Math Link

Dr. Edward Doolittle, a Mohawk Indian, was the first Indigenous person in Canada to obtain a PhD in Mathematics. Dr. Doolittle has taught at the First Nations University in Saskatchewan, and he is currently an Assistant Professor of Mathematics at the University of Regina. One of Dr. Doolittle's goals is to show his students how much fun mathematics can be. In addition to his academic interests, Dr. Doolittle also writes and performs comedy sketches for radio.

Reflect

How do you decide whether to solve an equation
by inspection or by systematic trial?
How might using a calculator affect your decision?
Give examples to illustrate your thinking.

Focus Solve equations by using a balance-scales model and verifying the solution.

Sometimes, systematic trial and inspection are not the best ways to solve an equation.

Balance scales can be used to *model* an equation. When pans are balanced, the mass in one pan is equal to the mass in the other pan.

We can write an equation to describe the masses in grams.
$20 = 10 + 5 + 5$

Explore

Use balance scales if they are available.
Otherwise, draw diagrams.

Here are some balance scales.
Some masses are known. Other masses are unknown.

Balance A

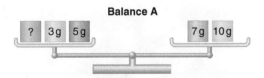

Balance B

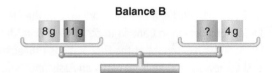

➤ The pans are balanced.
  For each balance scales:
  • Write an equation to represent the masses.
  • Find the value of the unknown mass.

➤ Make up your own balance-scales problem.
  Make sure the pans are balanced and one mass is unknown.
  Solve your problem.

Reflect & Share

Trade problems with another pair of classmates.
Compare strategies for finding the value of the unknown mass.

Connect

We can use a balance-scales model to solve an equation.

When the two pans of scales are balanced,
we can adjust each pan of the scales in the same way,
and the two pans will still be balanced.

➤ Consider these balance scales:

If we add 3 g to each pan,
the masses still balance.
$12 g + \mathbf{3 g} = 8 g + 4 g + \mathbf{3 g}$
$15 g = 15 g$

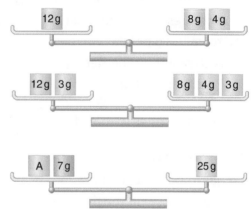

➤ Here is a balance-scales problem.
Mass A is an unknown mass.

If we remove 7 g from the left pan,
then Mass A is alone in that pan.
To keep the pans balanced,
we need to remove 7 g from the right pan too.
We replace 25 g with 7 g and 18 g;
then remove 7 g.

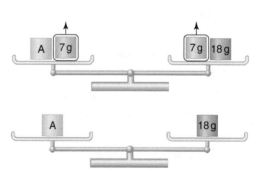

We are left with Mass A in the left pan
balancing 18 g in the right pan.
So, Mass A is 18 g.

We can *verify* the solution to this problem.
Replace Mass A with 18 g.
Then, in the left pan: $18 g + 7 g = 25 g$
And, in the right pan: 25 g
Since the masses are equal, the solution is correct.

**To verify means to check
the solution is correct.**

Example

A hockey team gets 2 points for a win,
1 point for a tie, and 0 points for a loss.
The Midland Tornadoes ended the season
with 28 points. They tied 6 games.
How many games did they win?
Write an equation you can use to solve the problem.
Use a model to solve the equation, then verify the solution.

A Solution

Let *w* represent the number of games won by the Midland Tornadoes.
So, 2*w* represents the number of points earned from wins.
The team has 6 points from ties.
It has 28 points altogether.
So, the equation is: $2w + 6 = 28$

Use balance scales to model the equation.

To get *w* on its own in one pan,
6 g has to be removed from the left pan.
So, select masses in the right pan
so that 6 g can be removed.
One way is to replace 28 g with 20 g + 6 g + 2 g.
Then, remove 6 g from each pan.

Two identical unknown masses
are left in the left pan.
20 g + 2 g = 22 g are left in the right pan.
Replace 22 g with two 11-g masses.

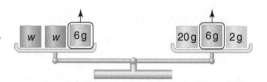

The two unknown masses balance
with two 11-g masses.
So, each unknown mass is 11 g.

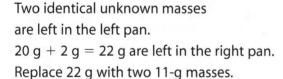

The Midland Tornadoes won 11 games.
Verify the solution by replacing
each unknown mass with 11 g.
11 + 11 + 6 = 28, so the solution is correct.

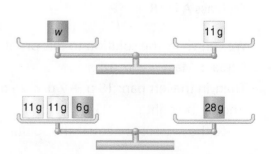

The examples on pages 227 and 228 show these ways in which we preserve balance and equality:

- We can *add* the same mass to each side.
- We can *subtract* the same mass from each side.
- We can *divide* each side into the same number of equal groups.

Later, we will show that:

- We can *multiply* each side by the same number by placing equal groups on each side that match the group already there.

Practice

1. Find the value of the unknown mass on each balance scales. Sketch the steps you used.

a)

b)

c)

d)

2. a) Sketch balance scales to represent each equation.
   b) Solve each equation. Verify the solution.
   - **i)** $x + 12 = 19$
   - **ii)** $x + 5 = 19$
   - **iii)** $4y = 12$
   - **iv)** $3m = 21$
   - **v)** $3k + 7 = 31$
   - **vi)** $2p + 12 = 54$

3. a) Write an equation for each sentence.
   b) Solve each equation. Verify the solution.
   - **i)** Five more than a number is 24.
   - **ii)** Eight more than a number is 32.
   - **iii)** Three times a number is 42.
   - **iv)** Five more than two times a number is 37.

4. The area of a rectangle is $A = bh$,
where b is the base of the rectangle and h is its height.
Use this formula for each rectangle below.
Substitute for A and b to get an equation.
Solve the equation for h to find the height.
Show the steps you used to get the answers.

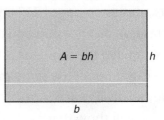

a)

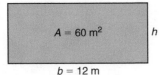

$A = 60$ m^2 h
$b = 12$ m

b)

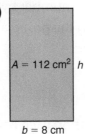

$A = 112$ cm^2 h
$b = 8$ cm

c)

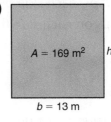

$A = 169$ m^2 h
$b = 13$ m

5. **Assessment Focus** Suppose the masses for balance scales
are only available in multiples of 5 g.
a) Sketch balance scales to represent this equation:
$x + 35 = 60$
b) Solve the equation. Verify the solution.
Show your work.

6. **Take It Further**
a) Write a problem that can be solved using
this equation: $x + 4 = 16$
b) How would your problem change if the equation were $x - 4 = 16$?
c) Solve the equations in parts a and b.
Show your steps.

7. **Take It Further** Write an equation you could
use to solve this problem.
Replace the □ in the number 5□7 with a digit
to make the number divisible by 9.

Reflect

Do you think you can always solve an equation using balance scales?
Justify your answer. Include an example.

Focus Use algebra tiles and inspection to solve equations involving integers.

Recall that 1 red unit tile and 1 yellow unit tile combine to model 0.
These two unit tiles form a zero pair.

−1   +1

The yellow variable tile represents a variable, such as x.

x

Explore

You will need algebra tiles.
Tyler had some gumdrops and jellybeans.
He traded 5 gumdrops for 5 jellybeans.
Tyler then had 9 gumdrops and 9 jellybeans.
How many gumdrops did he have to begin with?

Let g represent the number of gumdrops Tyler began with.
Write an equation you can use to solve for g.
Use tiles to represent the equation.
Use the tiles to solve the equation. Sketch the tiles you used.

Reflect & Share

Compare your equation with that of another pair of classmates.
Share your strategies for solving the equation using tiles.
How did you use zero pairs in your solutions?
Work together to find how many jellybeans Tyler began with.
Discuss your strategies for finding out.

Connect

Michaela has a collection of old pennies.
She sells 3 pennies to another collector.
Michaela then has 10 pennies left.
How many pennies did she have before she made the sale?

Let p represent the number of pennies Michaela
had before she made the sale.
The equation is: $p - 3 = 10$
One way to solve this equation is to use tiles.

Draw a vertical line in the centre of the page.
It represents the equal sign in the equation.
We arrange tiles on each side
of the line to represent an equation.
Recall that subtracting 3 is equivalent to adding -3.
So, we represent subtract 3 with 3 red unit tiles.

On the left side, put algebra
tiles to represent $p - 3$.

On the right side, put algebra
tiles to represent 10.

To isolate the variable tile, add 3
yellow unit tiles to make zero pairs.
Remove zero pairs.

Add 3 yellow unit tiles to this side, too,
to preserve the equality.

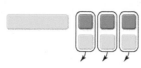

The tiles show the solution is $p = 13$.
Michaela had 13 old pennies before she made the sale.

Recall from Unit 1 that we can verify the solution by
replacing p with 13 yellow unit tiles.
Then:

becomes

Since there are now 10 yellow unit tiles on each side, the solution is correct.

Example

At 10 a.m., it was cold outside.
By 2 p.m., the temperature had risen 3°C to −6°C.
What was the temperature at 10 a.m.?

A Solution

Let t represent the temperature, in degrees Celsius, at 10 a.m.
After an increase of 3°C, the temperature was −6°C.
The equation is: $-6 = t + 3$

> The variable in an equation can be on the left side or the right side.

Add 3 red unit tiles to each side. Remove zero pairs.

9 red unit tiles equals one variable tile.

> We can verify the solution by replacing one yellow variable tile with 9 red unit tiles in the original equation.

The solution is $t = -9$.
At 10 a.m., the temperature was −9°C.

Another Solution

We can also solve equations involving integers by inspection.
To solve $-6 = t + 3$ by inspection:
We find a number which, when 3 is added to it, gives -6.
Think of moving 3 units to the right on a number line.
To arrive at -6, we would have to start at -9.
So $t = -9$.

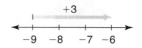

Practice

1. Use tiles to solve each equation.
Sketch the tiles you used.

a) $x + 4 = 8$ b) $3 + x = 10$ c) $12 = x + 2$

d) $x - 4 = 8$ e) $10 = x - 3$ f) $12 = x - 2$

2. Solve by inspection. Show your work.

a) $9 = n - 4$ b) $x + 6 = 8$ c) $2 = p - 5$

d) $x - 4 = -9$ e) $-8 = s + 6$ f) $x - 5 = -2$

3. Four less than a number is 13.
Let x represent the number.
Then, an equation is: $x - 4 = 13$
Solve the equation. What is the number?

4. Jody had some friends over to watch movies.
Six of her friends left after the first movie.
Five friends stayed to watch a second movie.
Write an equation you can use to find how many
of Jody's friends watched the first movie.
Solve the equation. Verify the solution.

5. Overnight, the temperature dropped 8°C to -3°C.
a) Write an equation you can solve
to find the original temperature.
b) Use tiles to solve the equation.
Sketch the tiles you used.

6. (Assessment Focus) Solve each equation using tiles, and by inspection. Verify each solution. Show your work.

   a) $x + 6 = 13$ **b)** $n - 6 = 13$

7. At the Jungle Safari mini-golf course, par on each hole is 5.
   A score of -1 means a player took 4 strokes to reach the hole.
   A score of $+2$ means a player took 7 strokes to reach the hole.
   Write an equation you can use to solve each problem below.
   Solve the equation. Show your work.

   a) On the seventh hole, Andy scored $+2$.
       His overall score was then $+4$.
       What was Andy's overall score after six holes?

   b) On the thirteenth hole, Bethany scored -2.
       Her overall score was then $+1$.
       What was Bethany's overall score
       after twelve holes?

   c) On the eighteenth hole,
       Koora reached the hole in one stroke.
       His overall score was then -2.
       What was Koora's overall score after seventeen holes?

> Par is the number of strokes a good golfer should take to reach the hole.

8. **Take It Further** Consider equations of the form $x + a = b$,
   where a and b are integers. Make up a problem that can
   be solved by an equation of this form in which:

   a) Both a and b are positive.
   b) Both a and b are negative.
   c) a is positive and b is negative.
   d) a is negative and b is positive.
   Solve each equation.
   Explain the method you used each time.

Reflect

How did your knowledge of adding and subtracting
integers help you in this lesson?

Mid-Unit Review

LESSON

6.1
1. Jaclyn went on a 4-day hiking trip.

a) For each problem, write an equation you can solve by inspection.

i) Jaclyn hiked 5 km the first day. After two days she had hiked 12 km. How far did she hike on the second day?

ii) Jaclyn hiked a total of 12 km on the third and fourth days. She hiked the same distance each day.
How far did she hike on each of these two days?

b) For each problem, write an equation you can solve by systematic trial.

i) Jaclyn counted squirrels. During the first three days, she counted a total of 67 squirrels. After four days, her total was 92 squirrels.
How many squirrels did she see on the fourth day?

ii) Jaclyn drank the same volume of water on each of the first three days. On the fourth day, she drank 8 cups of water. She drank 29 cups of water over the four days.
How many cups of water did she drink on each of the first three days?

6.2
2. a) Write an equation for each sentence.

b) Sketch balance scales to represent each equation.

c) Solve each equation. Verify the solution.

i) Nine more than a number is 17.

ii) Three times a number is 21.

iii) Seven more than two times a number is 19.

3. Andre's age is 14 more than twice Bill's age. Andre is 40 years old. How old is Bill? Write an equation you can use to solve this problem. Solve the equation using a balance-scales model.

6.3
4. a) Write an equation you can use to solve each problem.

b) Use tiles to solve each equation. Sketch the tiles you used. Verify each solution.

c) Solve each equation by inspection.

i) Eight years ago, Susanna was 7 years old.
How old is she now?

ii) The temperature dropped 6°C to −4°C. What was the original temperature?

iii) Hannah borrowed money. She paid back $7. Hannah still owes $5. How much money did she borrow?

6.4 Solving Equations Using Algebra

Focus   Solve a problem by solving an equation algebraically.

Explore

Solve this problem.
My mother's age is 4 more than 2 times my brother's age.
My mother is 46 years old. How old is my brother?

Reflect & Share

Discuss the strategies you used for finding
the brother's age with those of another pair of classmates.
Did you draw a picture? Did you use tiles? Did you use an equation?
If you did not use an equation, how could you represent
this problem with an equation?

Connect

Solving an equation *using algebra* is often the quickest way to find a solution,
especially if the equation involves large numbers.

Recall that when we solve an equation,
we find the value of the variable that makes the equation true.
That is, we find the value of the variable which,
when substituted into the equation,
makes the left side of the equation equal to the right side.

When we solve an equation using algebra, remember
the balance-scales model.
To preserve the equality, always perform the same operation
on both sides of the equation.

> Whatever you do to one
> side of an equation, you
> do to the other side too.

Example

Three more than two times a number is 27. What is the number?
a) Write an equation to represent this problem.
b) Solve the equation. Show the steps.
c) Verify the solution.

A Solution

a) Let n represent the number.

Then two times the number is: $2n$

And, three more than two times the number is: $2n + 3$

The equation is: $2n + 3 = 27$

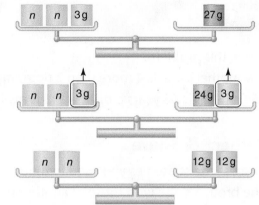

b) $2n + 3 = 27$

To isolate $2n$, subtract 3 from each side.

$2n + 3 - 3 = 27 - 3$

$2n = 24$

Divide each side by 2.

$$\frac{2n}{2} = \frac{24}{2}$$

$$n = 12$$

c) To verify the solution, substitute $n = 12$ into $2n + 3 = 27$.

Left side $= 2n + 3$ Right side $= 27$

$= 2(12) + 3$

$= 24 + 3$

$= 27$

Since the left side equals the right side, $n = 12$ is correct.

The number is 12.

> We could also verify by following the instructions in the *Example*.
> Start with 12.
> Multiply by 2: 24
> Add 3: 27

Practice

Solve each equation using algebra.

1. Solve each equation. Verify the solution.

   a) $x - 27 = 35$ **b)** $11x = 132$ **c)** $4x + 7 = 75$

2. Write, then solve, an equation to find each number. Verify the solution.

   a) Nineteen more than a number is 42.

   b) Ten more than three times a number is 25.

   c) Fifteen more than four times a number is 63.

3. Five years after Jari's age now doubles, he will be 27. How old is Jari now?

   a) Write an equation you can use to solve the problem.

   b) Solve the equation. Show the steps. How old is Jari?

   c) Verify the solution.

4. Jenny baby-sat on Saturday for $6/h. She was given a $3 bonus.
How many hours did Jenny baby-sit if she was paid $33?
  a) Write an equation you can use to solve the problem.
  b) Solve the equation. How many hours did Jenny baby-sit?
  c) Verify the solution.

5. In *x* weeks and 4 days, the movie *Math-Man IV* will be released.
The movie will be released in 25 days. Find the value of *x*.
  a) Write an equation you can use to solve the problem.
  b) Solve the equation. Verify the solution.

6. Look at the square and triangle on the right.
The sum of their perimeters is 56 cm.
The perimeter of the triangle is 24 cm.
What is the side length of the square?
  a) Write an equation you can use to find the side length of the square.
  b) Solve the equation. Verify the solution.

7. Assessment Focus Sunita has $72 in her savings account.
Each week she saves $24. When will Sunita have a total savings of $288?
  a) Write an equation you can use to solve the problem.
  b) Solve the equation. Show the steps.
     When will Sunita have $288 in her savings account?
  c) How can you check the answer?

8. Take It Further Use the information on the sign to the right.
  a) Write a problem that can be solved using an equation.
  b) Write the equation, then solve the problem.
  c) Show how you could solve the problem
     without writing an equation.

9. Take It Further The *n*th term of a number pattern is $9n + 1$.
What is the term number for each term value?
  a) 154   b) 118   c) 244

Reflect

What advice would you give someone who is having difficulty
solving equations using algebra?

Recall the methods you have used to solve an equation:
- using algebra tiles
- by inspection
- by systematic trial
- using a balance-scales model
- using algebra

Explore

Lila, Meeka, and Noel are playing darts.
Each player throws 3 darts at the board.
A player's score is the sum of the numbers
in the areas the darts land.
This picture shows a score of:
$8 + 20 + 2 = 30$

Write an equation for each problem.
Solve the equation using a method of your choice.

➤ Lila's first two darts scored a total of 12 points.
Lila scored 20 points in the round.
How many points did she score with her third dart?

➤ All three of Meeka's darts landed in the same area.
She scored 63 points. In which area did all her darts land?

➤ Noel's first two darts landed in the same area.
Her third dart was a bull's-eye, scoring 50 points.
She scored a total of 72 points.
In which area did her first two darts land?

Reflect & Share

Compare your equations with those of another pair of classmates.
Explain why you chose the method you did to solve them.
Use a different method to solve one of the equations.
Did this method work better for you? Why do you think so?

Connect

We can use any method to solve an equation, as long as the steps we take make sense, and the correct solution is found.

> **You can always check if the solution is correct by substituting the solution into the original equation.**

Example

In a basketball game between the Central City Cones and the Park Town Prisms, the lead changed sides many times. Write, then solve, an equation to solve each problem.

a) Early in the game, the Cones had one-half as many points as the Prisms. The Cones had 8 points. How many points did the Prisms have?

b) Near the end of the first half, the Cones were 12 points ahead of the Prisms. The Prisms had 39 points. How many points did the Cones have?

c) The Prisms scored 32 points in the fourth quarter. Twenty of these points were scored by foul shots and field goals. The rest of the points were scored by 3-point shots. How many 3-point shots did the Prisms make in the fourth quarter?

A Solution

a) Let p represent the number of points the Prisms had.
The Cones had one-half as many: $\frac{p}{2}$
The Cones had 8 points.
The equation is: $\frac{p}{2} = 8$
Solve using algebra.
Multiply each side by 2.

> **Another Strategy**
> We could use inspection to solve this equation.

$\frac{p}{2} \times 2 = 8 \times 2$

$\frac{2p}{2} = 16$

$p = 16$

The Prisms had 16 points.

b) Let d represent the number of points the Cones had.

The Prisms had 12 fewer points: $d - 12$

An equation is: $d - 12 = 39$

Solve using systematic trial. We choose a value for d and substitute.

Try $d = 50$. $d - 12 = 50 - 12$
                            $= 38$

38 is close. Choose a greater value for d.

Try $d = 51$. $d - 12 = 51 - 12$
                            $= 39$

This is correct. The Cones had 51 points.

> **Another Strategy**
> We could use algebra to solve this equation.

c) Let t represent the number of 3-point shots the Prisms made in the fourth quarter.

So, $3t$ represents the number of points scored by 3-point shots.

The equation is: $3t + 20 = 32$

Use a balance-scales model.

To isolate t, 20 g has to be removed from the left pan.

So, replace 32 g in the right pan with masses of 20 g and 12 g,

since 20 g + 12 g = 32 g. Then, remove 20 g from each pan.

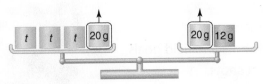

Three identical unknown masses remain in the left pan.

12 g remain in the right pan. Replace 12 g with three 4-g masses.

The three unknown masses balance with three 4-g masses.

So, each unknown mass is 4 g.

The Prisms made four 3-point shots in the fourth quarter.

In *Example,* part c, we can solve the equation using algebra to check.

$$3t + 20 = 32$$
$$3t + 20 - 20 = 32 - 20 \qquad \text{To isolate } 3t, \text{ subtract 20 from each side.}$$
$$3t = 12$$
$$\frac{3t}{3} = \frac{12}{3} \qquad \text{To isolate } t, \text{ divide each side by 3.}$$
$$t = 4$$

The algebraic solution and the balance-scales solution are the same.
It was much quicker to solve the equation using algebra.

Practice

Use algebra, systematic trial, inspection, algebra tiles,
or a balance-scales model to solve each equation.

1. Use algebra to solve each equation. Verify each solution.
   a) $\frac{x}{2} = 4$ b) $\frac{x}{3} = 7$ c) $\frac{x}{4} = 16$ d) $\frac{x}{5} = 10$

2. Which method would you choose to solve each equation?
   Explain your choice.
   Solve each equation using the method of your choice.
   a) $x + 5 = 12$ b) $x - 5 = 12$ c) $\frac{x}{6} = 9$ d) $x + 4 = -9$
   e) $4x = 36$ f) $16x = 112$ g) $4x + 2 = 30$ h) $8x + 17 = 105$

3. George and Mary collect friendship beads.
   George gave Mary 7 beads.
   Mary then had 21 beads.
   How many friendship beads did Mary have to start with?
   a) Write, then solve, an equation you can use to solve this problem.
   b) Verify the solution.

4. Jerome baked some cookies.
   He shared them among his eight friends.
   Each friend had 4 cookies.
   Write, then solve, an equation to find how
   many cookies Jerome baked.

5. Which method do you prefer to use to solve an equation?
   Explain. Give an example.

6. **Assessment Focus** Carla has 20 songs downloaded to her MP3 player.
Each month she downloads 8 additional songs.
After how many months will Carla have a total of 92 songs?
   a) Use an equation to solve the problem.
   b) Which method did you choose to solve the equation?
       Explain why you chose this method.

7. Write, then solve, an equation to answer each question.
Verify the solution.
Sheng sorted 37 cans.
   a) He divided the cans into 4 equal groups.
       He had 5 cans left over.
       How many cans were in each group?
   b) He divided the cans into 9 equal groups.
       He had 10 cans left over.
       How many cans were in each group?

8. Write, then solve, an equation to answer each question.
Verify the solution.
At Pascal's Pet Store, a 5-kg bag of dog food costs $10.
The 10-kg bag costs $15.
   a) Pascal sold $85 worth of dog food.
       He sold four 5-kg bags. How many 10-kg bags did he sell?
   b) Pascal sold $140 worth of dog food.
       He sold six 10-kg bags. How many 5-kg bags did he sell?

9. **Take It Further** Refer to the dart problem in *Explore*, page 240.
   a) Write two more problems using the given information.
       For each problem, write an equation you can use
       to solve your problem. Solve the equation.
       Use a different method for each equation.
   b) How could a player score 35 points with 3 darts?
       Find as many different ways as you can.

Reflect

Talk to a partner.
Tell how you choose the method you use to solve an equation.

Equation Baseball

Each game card is marked with a circled number to indicate how many bases you move for a correct answer. Each time a player crosses home plate on the way around the board, one run is awarded.

HOW TO PLAY THE GAME:

1. Shuffle the equation cards. Place them face down in the middle of the game board. Each player places a game piece on home plate.

2. Each player rolls the die. The player with the greatest number goes first. Play moves in a clockwise direction.

3. The first player turns over the top card for everyone to see. She solves the equation using a method of her choice. The other players check the answer. If the answer is correct, she moves the number of bases indicated by the circled number on the card and places the card in the discard pile. If the answer is incorrect, the card is placed in the discard pile.

4. The next player has a turn.

5. The player with the most runs when all cards have been used wins.

Decoding Word Problems

A word problem is a math question that has a story. Word problems often put math into real-world contexts.

The ability to read and understand word problems helps you connect math to the real world and solve more complicated problems.

Work with a partner. Compare these two questions.

Question 1   **Question 2**

Taylor has three more apples than Judy. Judy has six apples. How many apples does Taylor have?

$6 + 3$

List three reasons why the first question is more difficult than the second question.

A word problem can be challenging because it may not be obvious which math operations are needed ($+, -, \times, \div$) to solve it.

Work with a partner.
Solve each word problem below.

1. Julio has 36 photos of his favourite singing star.
He wants to arrange the photos in groups that have
equal numbers of rows and columns.
How many different arrangements can Julio make?
Show your work.

2. A rectangular garden is 100 m long and 44 m wide.
A fence encloses the garden.
The fence posts are 2 m apart.
Predict how many posts are needed.

3. A digital clock shows this time.
Seven minutes past 7 is a palindromic time.
List all the palindromic times between
noon and midnight.

What's the Question?

The *key words* in a word problem are the words
that tell you what to do.

Here are some common key words:

solve	explain
describe	predict
estimate	simplify
show your work	graph
find	list
compare	

Work with a partner. Discuss each question:
• What does each key word ask you to do?
• Which key words require an exact answer?
• Which key words tell you to show your thinking?

Unit Review

☑ We can solve equations:
- by inspection
- by systematic trial
- using a balance-scales model
- using algebra tiles
- using algebra

☑ To keep the balance of an equation, what you do to one side you must also do to the other side. This is called preserving equality.

What Should I Be Able to Do?

LESSON

6.1

1. Jan collects foreign stamps.
Her friend gives her 8 stamps.
Jan then has 21 stamps.
How many stamps did Jan have to start with? Let x represent the number of stamps.
Then, an equation is: $8 + x = 21$
Solve the equation.
Answer the question.

2. Write an equation for each sentence.
Solve each equation by inspection.
a) Five more than a number is 22.
b) Seven less than a number is 31.
c) Six times a number is 54.
d) A number divided by eight is 9.
e) Nine more than three times a number is 24.

3. Write an equation you can use to solve each problem. Solve each equation by systematic trial.
a) Ned spent $36 on a new shuttlecock racquet.
He then had $45 left.
How much money did Ned have before he bought the racquet?
b) Laurie sold 13 books for $208.
All books had the same price.
What was the price of each book?
c) Maurice sorts some dominoes.
He divides them into 15 groups, with 17 dominoes in each group.
How many dominoes does Maurice sort?

6.2 **4.** Write the equation that is represented by each balance scales. Solve the equation. Sketch the steps.

a)

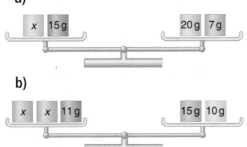

b)

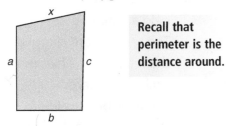

5. Look at the polygon below.

Find the value of x when:

Recall that perimeter is the distance around.

a) the perimeter of the polygon is 21 cm and $a = 5$ cm, $b = 3$ cm, and $c = 7$ cm
b) the perimeter of the polygon is 60 cm and $a = 15$ cm, $b = 11$ cm, and $c = 18$ cm

6. Jerry makes some photocopies. He pays 25¢ for a copy-card, plus 8¢ for each copy he makes. Jerry paid a total of 81¢. How many photocopies did Jerry make?

a) Write, then solve, an equation you can use to solve this problem. Show the steps.
b) Verify the solution.

6.3 **7.** Solve each equation using tiles, and by inspection. Verify each solution.

a) $x + 6 = 9$　　b) $n + 9 = 6$
c) $w - 6 = 9$　　d) $x - 9 = 6$

8. Adriano thinks of two numbers. When he adds 5 to the first number, the sum is -7. When he subtracts 5 from the second number, the difference is $+7$. What are the two numbers?

a) Write 2 equations you can use to solve this problem.
b) Use algebra tiles to solve the equations. Sketch the tiles you used.

9. Max spins the pointer on this spinner. He adds the number the pointer lands on to his previous total each time.

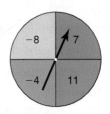

a) Write an equation you can use to solve each problem below.
b) Solve the equation using algebra tiles.
c) Verify each solution.
   i) Max gets -8 on his first spin. After his next spin, his total is $+3$. Which number did he get on his second spin?
   ii) After 3 spins, Max has a total of -1. Which number did he get on his third spin?

6.4 **10.** Sara collects 56 leaves for a science project. She collects the same number of each of 7 different types of leaves. How many of each type did Sara collect?

a) Write an equation you can use to solve the problem.

b) Solve the equation. Verify the solution.

11. Serena walks 400 m from home to school.
Serena is 140 m from school.
How far is Serena from home?

a) Write an equation you can use to solve the problem.

b) Solve the equation using algebra. Verify the solution.

12. The Grade 7 classes sold pins to raise money for charity.
They raised $228.
Each pin sold for $4.
How many pins did they sell?

a) Write an equation you can use to solve the problem.

b) Solve the equation using algebra. Verify the solution.

13. Write a problem that can be described by each equation below.
Solve each equation using algebra.
Explain the meaning of each answer.

a) $x + 15 = 34$ b) $7x = 49$

c) $\frac{x}{5} = 9$ d) $4x + 5 = 37$

6.5 **14.** Solve each equation using a method of your choice.
Explain why you chose each method.

a) $x + 12 = 24$ b) $x + 7 = -3$

c) $x - 18 = -15$ d) $4x = 28$

e) $\frac{x}{11} = 9$ f) $5x + 8 = 73$

15. Jaya has 25 hockey cards.
She has one more than 3 times the number of cards her brother has.
Write, then solve, an equation to find how many cards he has.

16. The school's sports teams held a banquet.
The teams were charged $125 for the rental of the hall, plus $12 for each meal served.
The total bill was $545.
How many people attended the banquet?

a) Write an equation you could use to solve the problem.

b) Solve your equation.

c) Verify the solution.

Practice Test

1. Solve each equation using a method of your choice.
   Explain your steps clearly.
   a) $x - 9 = -7$ b) $12p = 168$ c) $\frac{c}{7} = 9$ d) $7q + 11 = 102$

2. The area of a rectangle is $A = bh$,
   where b is the base of the rectangle and h is its height.
   The perimeter of a rectangle is $P = 2b + 2h$.
   - Write an equation you can use to solve each problem below.
   - Solve the equation. Verify the solution.
     a) What is the height of the rectangle when
        $b = 4$ cm and $A = 44$ cm²?
     b) What is the base of the rectangle when
        $h = 16$ cm and $P = 50$ cm?

   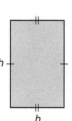

3. The formula $s = \frac{d}{t}$ relates average speed, s, distance, d, and time, t.
   Brad took part in a mini-marathon race.
   a) Brad jogged at an average speed of 5 km/h for 2 h.
      How far did he jog?
   b) Brad then rode his bike at an average speed of 16 km/h for 3 h.
      How far did he ride his bike?
   c) What distance was the race?
   Show how you used the formula to solve this problem.

4. Wapeka saves pennies. She has 12¢ in her jar at the start.
   Wapeka starts on January 1st. She saves 5¢ every day.
   - Write an equation you can use to solve each problem.
   - Solve the equation. Verify the solution.
     a) By which day had Wapeka saved a total of 47¢?
     b) By which day had Wapeka saved a total of $1.07?

5. Anoki is holding a skating party.
   The rental of the ice is $75, plus $3 per skater.
   a) Write an expression for the cost in dollars for 25 skaters.
   b) Suppose Anoki has a budget of $204. Write an equation
      you can solve to find how many people can skate.
      Solve the equation.

Suppose your older sister has bought an MP3 player.
She wants to download songs to play on it.
She asks for your help to find the best digital music club.

Part 1

1. Here are three plans from digital music clubs.
   Each plan includes 10 free downloaded songs per month.

   Songs4U: $20 per month, plus $3 per additional song
   YourHits: $30 per month, plus $2 per additional song
   Tops: $40 per month, plus $1 per additional song

   Copy and complete this table.

Number of Additional Songs per Month	3	6	9	12	15
Cost of Songs4U ($)					
Cost of YourHits ($)					
Cost of Tops ($)					

   What patterns do you see in the table?

2. Which club would you recommend if your sister plans to
   download 3 additional songs per month? 9 additional songs?
   15 additional songs? Explain your choice each time.

Part 2

3. **a)** For each plan, write an expression for the monthly cost of *n* additional songs.

b) Use each expression to find the total monthly cost for 10 additional songs for each club.

c) Suppose your sister can afford to spend $80 a month on downloading songs. Write an equation you can solve to find how many additional songs she can afford with each plan.

d) Solve each equation. Explain what each solution means.

Part 3

Write a paragraph to explain what decisions you made about choosing the best digital music club.

Reflect on Your Learning

You have learned different methods to solve an equation.
Which method do you prefer? Why?
Which method do you find most difficult?
What is it that makes this method so difficult?

UNIT

1

1. Use the divisibility rules to find the factors of each number.
   a) 120   b) 84   c) 216

2. Grace collects autographs of sports celebrities. She collected 7 autographs at the BC Open Golf Tournament. She then had 19 autographs. How many autographs did she have before the BC Open?
   a) Write an equation you can solve to find how many autographs Grace had before the BC Open.
   b) Solve the equation.
   c) Verify the solution.

2

3. Write the integer suggested by each situation below. Draw yellow or red tiles to model each integer.
   a) You walk down 8 stairs.
   b) You withdraw $10 from the bank.
   c) The temperature rises 9°C.

4. Use tiles to subtract.
   a) $(-9) - (-3)$ b) $(+9) - (-3)$
   c) $(+9) - (+3)$ d) $(-9) - (+3)$

3

5. Find a number between each pair of numbers.
   a) 1.6, 1.7   b) $\frac{6}{11}, \frac{7}{11}$
   c) $2\frac{1}{7}, \frac{16}{7}$ d) $2.7, 2\frac{4}{5}$

6. Find the area of a rectangular vegetable plot with base 10.8 m and height 5.2 m.

7. The regular price of a scooter is $89.99. The scooter is on sale for 20% off.
   a) What is the sale price of the scooter?
   b) There is 14% sales tax. What would a person pay for the scooter?

4

8. a) How many radii does a circle have?
   b) How many diameters does a circle have?

9. A DVD has diameter 12 cm.
   a) Calculate the circumference of the DVD. Give your answer to two decimal places.
   b) Estimate to check if your answer is reasonable.

10. Which has the greatest area? The least area?
    a) a rectangle with base 10 cm and height 5 cm
    b) a parallelogram with base 7 cm and height 8 cm
    c) a square with side length 7 cm

11. Zacharie has 50 m of plastic edging. He uses all the edging to enclose a circular garden. Find:
    a) the circumference of the garden
    b) the radius of the garden
    c) the area of the garden

12. Adele recorded the hair colours of all Grades 7 and 8 students in her school.

Hair Colour	Students
Black	60
Brown	20
Blonde	30
Red	10

a) Find the total number of students.
b) Write the number of students with each hair colour as a fraction of the total number of students.
c) Write each fraction as a percent.
d) Draw a circle graph to represent the data.

13. A cookie recipe calls for $\frac{3}{8}$ cup of brown sugar and $\frac{1}{3}$ cup of white sugar. How much sugar is needed altogether? Show your work.

14. Estimate, then add or subtract.
a) $\frac{3}{5} + \frac{1}{6}$ b) $\frac{5}{6} - \frac{5}{12}$
c) $\frac{2}{3} - \frac{1}{8}$ d) $\frac{1}{4} + \frac{2}{9}$

15. The Boudreau family started a trip with the gas gauge reading $\frac{3}{4}$ full. At the end of the trip, the gauge read $\frac{1}{8}$ full. What fraction of a tank of gas was used?

16. Add or subtract.
a) $5\frac{1}{6} + 3\frac{3}{4}$ b) $1\frac{3}{10} - \frac{2}{3}$
c) $1\frac{3}{5} + 3\frac{2}{3}$ d) $2\frac{5}{6} - 1\frac{5}{8}$

17. a) Sketch balance scales to represent each equation.
b) Solve each equation. Verify the solution.
   i) $s + 9 = 14$ ii) $s + 5 = 14$
   iii) $3s = 27$ iv) $3s + 5 = 23$

18. Use tiles to solve each equation. Sketch the tiles you used.
a) $x + 5 = 11$ b) $13 = x - 4$

19. Juan works as a counsellor at a summer camp. He is paid $7/h. He was given a $5 bonus for organizing the scavenger hunt. How many hours did Juan work if he was paid $250?
a) Write an equation you can use to solve the problem.
b) Solve the equation. How many hours did Juan work?

20. In a game of cards, black cards are worth +1 point each, and red cards are worth –1 point each. Four players are each dealt 13 cards per round, their scores are recorded, then the cards are shuffled. Write an equation you can use to solve each problem below. Solve the equation.
a) In Round Two, Shin was dealt 8 black cards and 5 red cards. His overall score was then +10. What was Shin's score after Round One?
b) In Round Two, Lucia was dealt 6 black cards and 7 red cards. Her overall score was then –4. What was Lucia's score after Round One?

Data Analysis

Many games involve probability. One game uses this spinner or a die labelled 1 to 6.

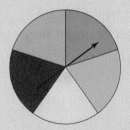 

You can choose to spin the pointer or roll the die.
You win if the pointer lands on red.
You win if you roll a 6.
Are you more likely to win if you spin the pointer or roll the die?
Why do you think so?

What You'll Learn

- Find the mean, mode, median, and range of a set of data.
- Determine the effect of an outlier on the mean, median, and mode.
- Determine the most appropriate average to report findings.
- Express probabilities as ratios, fractions, and percents.
- Identify the sample space for an experiment involving two independent events.
- Compare theoretical and experimental probability.

Why It's Important

- You see data and their interpretations in the media. You need to understand how to interpret these data.
- You need to be able to make sense of comments in the media relating to probability.

Key Words

- mean
- mode
- measure of central tendency
- average
- range
- median
- outlier
- chance
- impossible event
- certain event
- independent events
- tree diagram
- sample space

Focus Calculate the mean and mode for a set of data.

Questionnaires, experiments, databases, and the Internet are used to collect data. These collected data can be displayed in tables and graphs, which can be used to make predictions. In this lesson, you will learn ways to describe all the numbers in a data set.

Explore

You will need counters.
Three friends compared the time, in hours, they spent on the computer in one particular week.
Ali spent 5 h,
Bryn spent 9 h,
and Lynne spent 10 h.

Use counters to represent the time each person spent on the computer.
Find one number that best represents this time.

Reflect & Share

Share your findings with another pair of classmates.
How did you use counters to help you decide on the number?
Explain to your classmates why your number
best represents the data.

Connect

Allira surveyed 4 friends on the number of first cousins each has.
To find a number that best represents the number of cousins,
Allira used linking cubes.

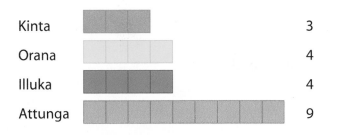

Kinta		3
Orana		4
Illuka		4
Attunga		9

The **mean** is a number that can represent the centre of a set of numbers.

➤ One way to find the mean is to rearrange the cubes to make rows of equal length.

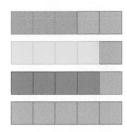

There are 5 cubes in each row.
The mean number of first cousins is 5.

When you make equal rows or columns,
the total number of cubes does not change.

➤ You can use the total number of cubes to calculate the mean.
The number of cubes in each row is 3, 4, 4, and 9.
Add these numbers: $3 + 4 + 4 + 9 = 20$
Then divide by the number of rows, 4: $20 \div 4 = 5$

The mean is 5.

The **mode** is the number that occurs most often.

➤ To find the mode, determine which number occurs most often.
In Allira's data, the number 4 occurs twice.
The mode is 4 cousins.
Two people have 4 cousins.

In a set of data,
there may be no
mode or there
may be more
than one mode.

Each of the mean and the mode is a **measure of central tendency**.
We say the word **average** to describe a measure of central tendency.
An average is a number that represents all numbers in a set.

Example

Here are Ira's practice times, in seconds, for the 100-m backstroke:
121, 117, 123, 115, 117, 119, 117, 120, 122
Find the mean and mode of these data.

A Solution

To find the mean practice time, add the practice times:
121 + 117 + 123 + 115 + 117 + 119 + 117 + 120 + 122 = 1071
Divide by the number of data, 9: 1071 ÷ 9 = 119
The mean practice time is 119 s.

The mode is the practice time that occurs most often.
117 occurs three times, so the mode practice time is 117 s.

Practice

1. Use linking cubes to find the mean of each set of data.
   a) 3, 4, 4, 5   b) 1, 7, 3, 3, 1   c) 2, 2, 6, 1, 3, 4

2. Calculate the mean of each set of data.
   a) 2, 4, 7, 4, 8, 9, 12, 4, 7, 3   b) 24, 34, 44, 31, 39, 32

3. Find the mode of each set of data in question 2.

4. Here are the weekly allowances for 10 Grade 7 students:
   $9, $11, $13, $15, $20, $10, $12, $15, $10, $15
   a) What is the mean allowance?
   b) What is the mode allowance?
   c) Suppose two allowances of $19 and $25 are added to the list.
      What is the new mean? What happens to the mode?

5. Here are the ages of video renters at *Movies A Must*
   during one particular hour: 10, 26, 18, 34, 64, 18, 21,
   32, 21, 54, 36, 16, 30, 18, 25, 69, 39, 24, 13, 22
   a) What is the mean age? The mode age?
   b) During another hour, the mode age of twelve video
      renters is 36. What might the ages of the renters be?
      Explain your answer.

6. Jordin Tootoo is the first Inuk athlete to play in the National Hockey League. On October 9, 2003, he played his first game for the Nashville Predators. This table shows Jordin's statistics when he played junior hockey for the Brandon Wheat Kings.

Jordin Tootoo's Scoring Records 1999-2003				
Year	Games Played	Goals	Assists	Points
1999-2000	45	6	10	16
2000-2001	60	20	28	48
2001-2002	64	32	39	71
2002-2003	51	35	39	74

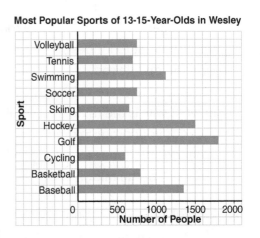

Find the mean and mode for each set of data.
a) Games Played
b) Goals
c) Assists
d) Points

7. Assessment Focus The graph shows the most popular sports of 13–15-year-olds in Wesley.
a) Which sports are equally popular?
b) How could you use the bar graph to find the mode?
Explain and show your work.
c) Calculate the mean.
Use estimated values from the graph.

Most Popular Sports of 13-15-Year-Olds in Wesley

8. Take It Further A data set has 6 numbers.
Four of the numbers are: 6, 3, 7, 9
Find the other two numbers in each case.
a) The mean is 6.
b) The mode is 3 and the mean is 6.
Find as many different answers as you can.

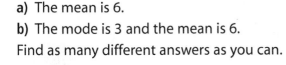

What is the difference between mean and mode?
Create a set of data to explain.

Median and Range

Focus Find the median and the range of a set of data.

The graph shows the number of tubes of hair gel used by each of 5 students in one particular month.

How many tubes of gel did each student use?
What is the mean number of tubes used?
The mode number?
How did you find the mean and the mode?

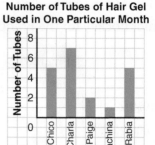

Number of Tubes of Hair Gel Used in One Particular Month

Explore

Your teacher will give you a bag of Cuisenaire rods.
You will need a ruler.

➤ Without looking, each person takes
3 rods from the bag.
Work together to arrange the 9 rods
from shortest to longest.
Find the middle rod.
How many rods are to its right? To its left?
In what way is the middle rod typical of
the rods your group picked?
What do you notice about the rods to the left
and right of the middle rod?

➤ Each of you takes 1 more rod from the bag.
Place them among the ordered rods in the appropriate places.
Is there a middle rod now? Explain.
Sketch the rods.
Below each rod in your sketch, write its length.
How could you use the lengths to find a "middle" length?
How is the middle length typical of the rods in your sketch?

Reflect & Share

Is it possible to have two different sets of rods with the same middle length?
Share your results with other groups to find out.

The **median** of a data set is the middle number when the data are arranged in order.

➤ There are 11 Grade 7 students in Ms. Shim's combined Grades 6 and 7 class.
To find the median mark on the last science test, she listed their marks from greatest to least:

95, 92, 87, 85, 80, 78, 76, 73, 70, 66, 54

The middle number is 78.
There are 5 marks greater than 78, and 5 marks less than 78.
The median mark is 78.

> When there is an odd number of data, the median is the middle number.

➤ Another Grade 7 student transfers to Ms. Shim's class.
He writes the same test and receives a mark of 72.

To find the new median, the teacher includes his mark in the ordered list:

95, 92, 87, 85, 80, 78, 76, 73, 72, 70, 66, 54

There are two middle numbers, 78 and 76.
There are 5 marks greater than 78, and 5 marks less than 76.
The median is the mean of the 2 middle numbers:
$(78 + 76) \div 2 = 77$
The median mark is now 77.

> When there is an even number of data, the median is the mean of the two middle numbers.

> The median is also a measure of central tendency, or an average.

➤ Now that the marks are arranged in order, we can easily find the range.
The **range** of a data set tells how spread out the data are.
It is the difference between the greatest and least numbers.
To find the range of the marks on the science test, subtract the least mark from the greatest mark:
$95 - 54 = 41$
The range of the marks is 41.

> When there is an even number of data, the median might *not* be one of the numbers in the data set.

Example

The hourly wages, in dollars, of 10 workers are: 8, 8, 8, 8, 9, 9, 9, 11, 12, 20

Find:

a) the mean **b)** the mode **c)** the median **d)** the range

How does each average relate to the data?

A Solution

a) Mean wage:

Add: $8 + 8 + 8 + 8 + 9 + 9 + 9 + 11 + 12 + 20 = 102$
Divide by the number of workers, 10: $102 \div 10 = 10.2$
The mean wage is $10.20.
Three workers have a wage greater than the mean and 7 workers
have a wage less than the mean.

b) Mode wage:

8, 8, 8, 8, 9, 9, 9, 11, 12, 20
The mode wage is $8. It occurs 4 times.
This is the least wage; that is, 6 workers have a wage greater than the mode.

c) Median wage:

List the 10 wages in order from least to greatest:
8, 8, 8, 8, **9, 9**, 9, 11, 12, 20
The median wage is the mean of the 5th and 6th wages.
Both the 5th and 6th wages are 9.
The median wage is $9.
There are 3 wages above the median and 4 wages below the median.

d) Range:

8, 8, 8, 8, 9, 9, 9, 11, 12, **20**
Subtract the least wage from the greatest wage: $20 - 8 = 12$
The range of the wages is $12.

Practice

1. Find the median and the range of each set of data.

    a) 85, 80, 100, 90, 85, 95, 90

    b) 12 kg, 61 kg, 85 kg, 52 kg, 19 kg, 15 kg, 21 kg, 30 kg

2. The Grade 7 students in two combined Grades 6 and 7 classes
wrote the same quiz, marked out of 15.
Here are the results:
Class A: 8, 9, 9, 12, 12, 13, 13, 14, 15, 15
Class B: 10, 10, 11, 11, 12, 12, 13, 13, 14, 14
   a) Find the median mark for each class.
   b) Find the range of each set of marks.
   c) Which class do you think is doing better? Explain.

3. a) Find the mean, median, and mode for each data set.
   i) 4, 5, 7, 8, 11   **ii)** 50, 55, 65, 70, 70, 50
   iii) 7, 63, 71, 68, 71   **iv)** 6, 13, 13, 13, 20
   b) Which data sets have:
   • the same values for the mean and median?
     What do you notice about the numbers in each set?
   • the same values for the mean, median, and mode?
     What do you notice about the numbers in each set?
   • different values for the mean, median, and mode?
     What do you notice about the numbers in each set?

4. Assessment Focus Write two different data sets with 6 numbers, so that:
   a) The mode is 100. The median and the mean are equal.
   b) The mode is 100. The mean is less than the median.
   Show your work.

5. a) The median height of ten 12-year-old girls is 158 cm.
     What might the heights be? How do you know?
   b) The mode height of ten 12-year-old boys is 163 cm.
     What might the heights be? How do you know?

6. Jamal was training for a 400-m race. His times, in seconds,
for the first five races were: 120, 118, 138, 124, 118
   a) Find the median and mode times.
   b) Jamal wants his median time after 6 races to be 121 s.
     What time must he get in his 6th race? Show your work.
   c) Suppose Jamal fell during one race and recorded a time of 210 s.
     Which of the mean, median, and mode would be most affected?
     Explain.

7. In 2005, the Edmonton Miners hosted The Minto Cup Junior A Lacrosse Championship. Here are the 2005 statistics, as of June 30, 2005, for 10 players on the team.

Player	Games	Goals	Assists	Points	Penalty Minutes
Jeremy Boyd	13	2	8	10	54
Dan Claffey	11	3	11	14	33
Dalen Crouse	11	10	10	20	6
Andrew Dixon	15	4	5	10	47
Dan Hartzell	11	5	21	26	8
Cole Howell	12	21	13	34	0
Aiden Inglis	12	3	4	7	23
Ryan Polny	17	7	14	21	2
Chris Schmidt	5	8	4	12	2
Neil Tichkowsky	17	34	19	53	8

a) Calculate the mean, the median, and the mode of each set of data.

b) Make up a question about the mean, the median, or the mode that can be answered using these data. Answer your question.

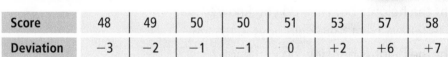

8. **Take It Further** This is how Edward calculated the mean of these data.

48, 49, 50, 50, 51, 53, 57, 58

Estimated mean is 51.

Score	48	49	50	50	51	53	57	58
Deviation	−3	−2	−1	−1	0	+2	+6	+7

$$\text{Mean} = 51 + \frac{(-3) + (-2) + (-1) + (-1) + 0 + 2 + 6 + 7}{8}$$
$$= 52$$

Check that Edward's answer is correct. How does his method work?

Reflect

A median is the strip of land or concrete barrier separating lanes of highway traffic travelling in opposite directions.
How is this meaning similar to its meaning in math?

7.3 The Effects of Outliers on Average

Focus Understand how mean, median, and mode are affected by outliers.

Explore

Students in a Grade 7 class measured their pulse rates.
Here are their results in beats per minute:
97, 69, 83, 66, 78, 8, 55, 82, 47, 52, 67, 76, 84,
64, 72, 80, 72, 70, 69, 80, 66, 60, 72, 88, 88

➤ Calculate the mean, median, and mode for these data.
➤ Are there any numbers that are significantly different from the rest?
  If so, remove them.
  Calculate the mean, median, and mode again.
  Explain how the three averages are affected.

Reflect & Share

Compare your results with those of another pair of classmates.
How did you decide which numbers were significantly different?
Why do you think they are so different?

Connect

A number in a set of data that is significantly different from
the other numbers is called an **outlier**.

An outlier is much greater than or much less than most of the
numbers in the data set.

Outliers sometimes occur as a result of error in measurement or recording.
In these cases, outliers should be ignored.

Sometimes an outlier is an important piece of information that should not
be ignored. For example, if one student does much better or much worse
than the rest of the class on a test.

Outliers may not always be obvious.
Identifying outliers is then a matter of choice.

Example

Here are the marks out of 100 on an English test for students in a Grade 7 class:

21, 23, 24, 24, 27, 29, 29, 29, 32, 37, 37, 38, 39,

40, 50, 50, 51, 54, 56, 57, 58, 59, 61, 71, 80, 99

a) How many students were in the class? How do you know?

b) What is the outlier? Explain your choice.

c) Calculate the mean, median, and mode.

d) Calculate the mean, median, and mode without the outlier.
   What do you notice?

e) Should the outlier be used when reporting the average test mark? Explain.

A Solution

a) Count the number of marks to find the number of students in the class.
   There are 26 students.

b) There is only one number, 99, that is significantly different.
   The outlier is 99.
   The difference between the outlier and the nearest mark is $99 - 80 = 19$.
   This difference is much greater than that between other pairs of adjacent marks.

c) There are 26 marks. To find the mean mark, add the marks:
   $21 + 23 + 24 + 24 + 27 + 29 + 29 + 29 + 32 + 37 + 37 + 38 + 39 +$
   $40 + 50 + 50 + 51 + 54 + 56 + 57 + 58 + 59 + 61 + 71 + 80 + 99 = 1175$
   Divide the total by the number of marks, 26: $1175 \div 26 \doteq 45.2$
   The answer is written to the nearest tenth.
   The mean mark is about 45.2.
   The median mark is the mean of the 13th and 14th marks.
   The 13th mark is 39. The 14th mark is 40.
   So, the median is: $\frac{39 + 40}{2} = \frac{79}{2} = 39.5$
   The mode is the mark that occurs most often. This is 29.

d) Without the outlier, there are 25 marks and the sum of the marks is: $1175 - 99 = 1076$
   The mean is: $1076 \div 25 = 43.04$
   The median is the 13th mark: 39
   The mode is 29.
   When the outlier was removed, the mean and median decreased.
   The mode remained the same.

e) The outlier should be used when reporting the average test mark.
   To understand how the class is performing, all test marks should be included.

Practice

1. This set of data represents the waiting time, in minutes, at a fast-food restaurant:

5, 5, 5, 6, 5, 7, 0, 5, 1, 7, 7, 5, 6, 5, 5, 5, 8, 5, 0, 5, 4, 5, 2, 7, 9

a) Calculate the mean, median, and mode.

b) Identify the outliers. Explain your choice.

c) Calculate the mean, median, and mode without the outliers. How is each average affected when the outliers are not included?

> Remember to arrange the data in order before finding the median.

2. Bryan recorded the time he spent on the school bus each day for one month. Here are the times, in minutes:

15, 21, 15, 15, 18, 19, 14, 20, 95, 18, 21, 14, 15, 20, 16, 14, 22, 21, 15, 19

a) Calculate the mean, median, and mode times.

b) Identify the outlier. How can you explain this time?

c) Calculate the mean, median, and mode times without the outlier. How is each average affected when the outlier is not included?

d) A classmate asks Bryan, "What is the average time you spend on the bus each day?" How should Bryan answer? Give reasons.

3. A clothing store carries pant sizes 28 to 46. A sales clerk records the sizes sold during her 6-h shift:

28, 36, 32, 32, 34, 4, 46, 44, 42, 38, 36, 36, 40, 32, 36

a) Calculate the mean, median, and mode sizes.

b) Is there an outlier? If so, why do you think it is an outlier?

c) Calculate the mean, median, and mode sizes without the outlier. How is each average affected when the outlier is not included?

d) Should the outlier be used when the sales clerk reports the average pant size sold during her shift? Explain your thinking.

4. Here are the science test marks out of 100 for the Grade 7 students in a combined-grades class:

0, 66, 65, 72, 78, 93, 82, 68, 64, 90, 65, 68

a) Calculate the mean, median, and mode marks.

b) Identify the outlier. How might you explain this mark?

c) Calculate the mean, median, and mode marks without the outlier. How is each average affected when the outlier is not included?

d) Should the outlier be used when reporting the average test mark? Explain.

5. a) Give an example of a situation in which outliers would not be used in reporting the averages. Explain why they would not be included.

b) Give an example of a situation in which outliers would be used in reporting the averages. Explain why they would be included.

6. Assessment Focus A Grade 7 class wanted to find out if a TV advertisement was true. The ad claimed that *Full of Raisins* cereal guaranteed an average of 23 raisins per cup of cereal. Each pair of students tested one box of cereal. Each box contained 20 cups of cereal. The number of raisins in each cup was counted.

a) Assume the advertisement is true. How many raisins should there be in 1 box of cereal?

b) Here are the results for the numbers of raisins in 15 boxes of cereal:

473, 485, 441, 437, 489, 471, 400, 453, 465, 413, 499, 428, 419, 477, 467

  i) Calculate the mean, median, and mode numbers of raisins.

  ii) Identify the outliers. Explain your choice.

  iii) Calculate the mean, median, and mode without the outliers. How do the outliers affect the mean?

  iv) Should the outliers be used when reporting the average number of raisins? Explain.

  v) Was the advertisement true? Justify your answer.

7. Take It Further Here is a set of data: 2, 3, 5, 5, 7, 8
An outlier has been removed.

a) Calculate the mean, median, and mode without the outlier.

b) The outlier is returned to the set. The averages become:

Mean: 7   Median: 5   Mode: 5

What is the outlier? Show your work.

Reflect

Your friend is having difficulty recognizing outliers in a data set. What advice would you give your friend?

7.4 Applications of Averages

Focus Understand which average best describes a set of data.

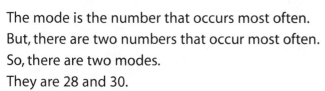

Explore

Record on the board how many siblings you have.
Use the class data.
Find the mean, the median, and the mode.
Find the range.

Reflect & Share

With a classmate, discuss which measure
best describes the average number of siblings.

Connect

A clothing store sold jeans in these sizes in one day:

28 30 28 26 30 32 28 32 26 28 34 38 36 30 34 32 30

To calculate the mean jeans size sold, add the sizes, then divide by
the number of jeans sold.

$$\text{Mean} = \frac{28 + 30 + 28 + 26 + 30 + 32 + 28 + 32 + 26 + 28 + 34 + 38 + 36 + 30 + 34 + 32 + 30}{17}$$

$$= \frac{522}{17}$$

$$\doteq 30.7$$

The mean size is approximately 30.7.

To calculate the median, order the jeans sold from least size to
greatest size. There are 17 numbers.
The middle number is the median. The middle number is the 9th.

26, 26, 28, 28, 28, 28, 30, 30, **30**, 30, 32, 32, 32, 34, 34, 36, 38

The median size is 30.

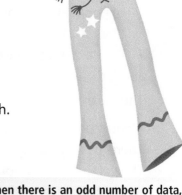

The mode is the number that occurs most often.
But, there are two numbers that occur most often.
So, there are two modes.
They are 28 and 30.
So, the mode sizes are 28 and 30.

> When there is an odd number of data,
> to find the middle number: Add 1 to
> the number of data, then divide by 2.
> This gives the position of the middle
> number. For example: $\frac{17 + 1}{2} = \frac{18}{2} = 9$;
> the middle number is the 9th.

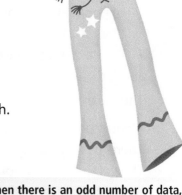

In this situation, the mean, 30.7, is of little use.
The mean does not represent a size.

The median, 30, shows about one-half of the customers bought jeans of size 30 or smaller, and about one-half of the customers bought jeans of size 30 or larger.

The modes, 28 and 30, tell which sizes are purchased more often.
The mode is most useful to the storeowner.
He may use the mode to order extra stock of the most popular sizes.

Example

A bookstore has 15 books in its young adult section.
There are 5 different prices.
This table shows the number of books at each price.
a) Find the mean, median, and mode prices.
b) Which measure best represents the
   average price of a young adult book?
c) What is the range of the prices?

Young Adult Books	
Price ($)	Number of Books
8.99	3
9.99	5
13.99	5
32.99	1
37.99	1

A Solution

Make a list of the prices, in dollars:
8.99, 8.99, 8.99, 9.99, 9.99, 9.99, 9.99, 9.99, 13.99,
13.99, 13.99, 13.99, 13.99, 32.99, 37.99
a) Mean price:
   • Multiply each price by the total number
     of books at that price, then add the prices.
     $(8.99 \times 3) + (9.99 \times 5) + (13.99 \times 5) + 32.99 + 37.99 = 217.85$
   • Divide the total price by the total number of books: 15
     $\frac{217.85}{15} \doteq 14.52$, to two decimal places
   The mean price is approximately $14.52.

   Median price:
   There are 15 books.
   The list shows the books in order from least price
   to greatest price.
   The median price is the 8th price. The 8th price is $9.99.
   The median price is $9.99.

   Mode price:
   There are two mode prices. They are $9.99 and $13.99.

b) The mean price is not charged for any of the books.
Only two books cost more than the mean of $14.52.

There are two mode prices.
One mode, $9.99, is the same as the median price.

One-half the books cost the median price or less.
One-half cost more. So, the median price, $9.99, best represents the average price
of a young adult book at the store.

c) For the range, subtract the lowest price from the highest price:
37.99 − 8.99 = 29.00
The range of prices is $29.00.

The mean is usually the best average when no numbers in the
data set are significantly different from the other numbers.

The median is usually the best average when there are numbers
in the data set that are significantly different.

The mode is usually the best average when the data represent measures,
such as shoe sizes or clothing sizes.
A store needs to restock the sizes that sell most often.

Practice

1. The daily high temperatures for one week at Clearwater
Harbour were: 27°C, 31°C, 23°C, 25°C, 28°C, 23°C, 28°C
   a) Find the mean, median, and mode for these data.
   b) Which average do you think best describes the daily high
   temperature at Clearwater Harbour that week? Explain.
   c) The weather channel reported the average temperature for
   Clearwater Harbour that week was 23°C. Is this correct? Explain.

2. Caitlin received these test marks in each subject.
   a) Find the mean, median, and mode mark for
   each subject.
   b) Explain what information each average gives.
   c) Which subject do you think Caitlin is best at?
   Worst at? Explain your reasoning.

Caitlin's Marks						
Math	85	69	92	55	68	75	78
Music	72	81	50	69	81	96	92
French	68	74	82	80	76	67	74

3. The table shows the tips earned by five waiters and waitresses during two weeks in December.

a) Calculate the mean, median, and mode tips for each week.

b) Calculate the mean, median, and mode tips for the two-week period.

c) Compare your answers in parts a and b. Which are the same? Which are different? Explain why.

d) Explain which average best represents the tips earned during the two weeks.

Weekly Tips Earned ($)		
Waiter	Week 1	Week 2
James	1150	600
Kyrra	700	725
Tamara	800	775
Jacob	875	860
George	600	1165

4. A small engineering company has an owner and 5 employees. This table shows their salaries.

a) Calculate the mean, median, and mode annual salaries.

b) What is the range of the annual salaries?

c) Which measure would you use to describe the average annual salary in each case? Explain.

   i) You want to attract a new employee.

   ii) You want to suggest the company does not pay its employees well.

Company Salaries	
Position	Annual Salary ($)
Owner	130 000
Manager	90 000
2 Engineers	50 000
Receptionist	28 000
Secretary	28 000

5. Is each conclusion correct? Explain your reasoning.

a) The mean cost of a medium pizza is $10.
So, the prices of three medium pizzas could be $9, $10, and $11.

b) The number of raisins in each of 30 cookies was counted.
The mean number of raisins was 15.
So, in 10 cookies, there would be a total of 150 raisins.

6. **Assessment Focus** In each case, which average do you think is most useful: the mean, median, or mode? Justify your answer.

a) A storeowner wants to know which sweater sizes she should order.
Last week she sold 5 small, 15 medium, 6 large, and 2 X-large sweaters.

b) Five of Robbie's friends said their weekly allowances are:
$10, $13, $15, $11, and $10.
Robbie wants to convince his parents to increase his allowance.

c) Tina wants to know if her math mark was in the top half or bottom half of the class.

7. A quality control inspector randomly selects
boxes of crackers from the production line.
She measures their masses.
On one day she selects 15 boxes, and records these data:
- 6 boxes: 405 g each
- 2 boxes: 395 g each
- 4 boxes: 390 g each
- 2 boxes: 385 g each
- 1 box: 380 g

a) Calculate the mean, median, and mode masses.

b) What is the range of the masses?

c) For the shipment of crackers to be acceptable,
the average mass must be at least 398 g.
Which average would you use to describe
this shipment to make it acceptable? Explain.

8. Take It Further Andrew has these marks:
English 82%, French 75%, Art 78%, Science 80%

a) What mark will Andrew need in math if he wants his mean
mark in these 5 subjects to be each percent?

   i) 80%   **ii)** 81%   **iii)** 82%

b) Is it possible for Andrew to get
a mean mark of 84% or higher?
Justify your answer.

9. Take It Further Celia received a mean mark of 80%
in her first three exams.
She then had 94% on her next exam.
Celia stated that her overall mean mark was 87%
because the mean of 80 and 94 is 87.
Is Celia's reasoning correct? Explain.

Reflect

Use your answers from *Practice*. Describe a situation for each case.

a) The mean is the best average.

b) The median is the best average.

c) The mode is the best average.

Justify your choices.

Using Spreadsheets to Investigate Averages

Focus | Investigate averages using a spreadsheet.

You can use spreadsheet software to find the mean, median, and mode of a set of data.
A spreadsheet program allows us to calculate the averages for large sets of data values quickly and efficiently.
You can also use the software to see how these averages are affected by outliers.

Here are the heights, in centimetres, of all Grade 7 students who were on the school track team: 164, 131, 172, 120, 175, 168, 146, 176, 175, 173, 155, 170, 172, 160, 168, 178, 174, 184, 189

> **In some spreadsheet software, the mean is referred to as the average.**

Use spreadsheet software.

➤ Input the data into a column of the spreadsheet.

➤ Use the statistical functions of your software to find the mean, median, and mode. Use the Help menu if you have any difficulties.

A20		ƒ	=AVERAGE(A1:A19)	
	A	B	C	D	E
1	164				
2	131				
3	172				
4	120				
5	175				
6	168				
7	146				
8	176				
9	175				
10	173				
11	155				
12	170				
13	172				
14	160				
15	168				
16	178				
17	174				
18	184				
19	189				
20	165.7895				
21	172				
22	172				

Median   Mode

➤ Investigate the effect of an outlier on the mean, median, and mode.
Delete 120.
What happens to the mean?
Median? Mode? Explain.

➤ Suppose one member of the track team with height 155 cm is replaced by a student with height 186 cm.
How does this substitution affect the mean, median, and mode? Explain.

	X ✓ ƒ	186	
	A	B	C
1	164		
2	131		
3	172		
4	120		
5	175		
6	168		
7	146		
8	176		
9	175		
10	173		
11	186		
12	170		

Notice that when you add or remove data values, the averages change to reflect the adjustments.

✓ Check

1. Enter these data into a spreadsheet.

They are the donations, in dollars, that were made to a Toy Wish Fund.

5, 2, 3, 9, 10, 5, 2, 8, 7, 15, 14, 17, 28, 30, 16, 19, 4, 7, 9, 11, 25, 30,

32, 15, 27, 18, 9, 10, 16, 22, 34, 19, 25, 18, 20, 17, 9, 10, 15, 35

a) Find the mean, median, and mode.

b) Add some outliers to your spreadsheet.

State the values you added.

How do the new mean, median, and mode compare to

their original values? Explain.

2. Enter these data into a spreadsheet.

They are purchases, in dollars, made by customers at a grocery store.

55.40, 48.26, 28.31, 14.12, 88.90, 34.45, 51.02, 71.87, 105.12, 10.19,

74.44, 29.05, 43.56, 90.66, 23.00, 60.52, 43.17, 28.49, 67.03, 16.18,

76.05, 45.68, 22.76, 36.73, 39.92, 112.48, 81.21, 56.73, 47.19, 34.45

a) Find the mean, median, and mode.

b) Add some outliers to your spreadsheet.

State the values you added.

How do the new mean, median, and mode compare to

their original values? Explain.

3. Enter these data into a spreadsheet.

They are the number of ice-cream bars sold at the community centre

each day in the month of July.

101, 112, 127, 96, 132, 125, 116, 97, 124, 136, 123, 113, 78, 102, 118, 130,

87, 108, 114, 99, 126, 86, 94, 117, 121, 107, 122, 119, 111, 105, 93

Find the mean, median, and mode.

What happens when you try to find the mode? Explain.

4. Repeat question 1 parts a and b.

This time, enter data you find in the newspaper or on the Internet.

Reflect

List some advantages of using a spreadsheet to find the mean, median,
and mode of a set of data. What disadvantages can you think of?

Mid-Unit Review

LESSON

7.1
7.2

1. Here are the heights, in centimetres, of the students in a Grade 7 class:

162, 154, 166, 159, 170, 168, 158, 162, 172, 166, 157, 170, 171, 165, 162, 170, 153, 167, 164, 169, 167, 173, 170

a) Find the mean, median, and mode heights.

b) What is the range of the heights?

2. The mean of five numbers is 20.
The median is 23.
What might the numbers be?
Find 2 different sets of data.

7.3

3. The cost of hotel rooms at *Stay in Comfort* range from $49 to $229 per night. Here are the rates charged, in dollars, for one particular night:

70, 75, 85, 65, 75, 90, 70, 75, 60, 80, 95, 85, 75, 20, 65, 229

a) Calculate the mean, median, and mode costs.

b) Identify the outliers.
How can you explain these costs?

c) Calculate the mean, median, and mode costs without the outliers.
How is each average affected when the outliers are not included?

d) Should the outliers be used when reporting the average cost of a hotel room? Explain.

7.4 **4.** A quality control inspector measures the masses of boxes of raisins. He wants to know if the average mass of a box of raisins is 100 g. The inspector randomly chooses boxes of raisins.
The masses, in grams, are:

99.1, 101.7, 99.8, 98.9, 100.8, 100.3, 98.3, 100.0, 97.8, 97.6, 98.5, 101.7, 100.2, 100.2, 99.4, 100.3, 98.8, 102.0, 100.3, 98.0, 99.4, 99.0, 98.1, 101.8, 99.8, 101.3, 100.5, 100.7, 98.7, 100.3, 99.3, 102.5

a) Calculate the mean, median, and mode masses.

b) For the shipment to be approved, the average mass of a box of raisins must be at least 100 g. Which average could someone use to describe this shipment to get it approved? Explain.

5. Is each conclusion true or false? Explain.

a) The mode number of books read last month by students in James' class is 5. Therefore, most of the students read 5 books.

b) A random sample of 100 people had a mean income of $35 000. Therefore, a random sample of 200 people would have a mean income of $70 000.

Focus   Express probabilities as ratios, fractions, and percents.

When the outcomes of an experiment are equally likely, the probability of an event occurring is:

Number of outcomes favourable to that event
<u>   Number of possible outcomes</u>

Explore

At the pet store, Mei buys 100 biscuits
for her dog, Ping-Ping.
She buys 75 beef-flavoured biscuits, 15 cheese-flavoured,
and 10 chicken-flavoured.
The clerk puts them all in one bag.
When she gets home, Mei shakes the bag
and pulls out one biscuit.

- What is the probability that Mei pulls out
  a cheese-flavoured biscuit from the bag?
- How many different ways could you write
  this probability?
- What is the probability of pulling out a beef-flavoured
  biscuit? A chicken-flavoured biscuit?
  Write each probability 3 different ways.
- What is the probability of pulling out a vegetable-flavoured biscuit?
- What is the probability of pulling out a flavoured biscuit?

Reflect & Share

Compare your results with those of another pair of classmates.
How many different ways did you write a probability?
Are all of the ways equivalent?
How do you know?
What is the probability of an event that always occurs?
An event that never occurs?

Connect

A probability can be written as a ratio, a fraction, and a percent.

Sam buys a box of different flavours of food for his cat.
In a box, there are 14 packets of fish flavour, 2 of chicken flavour, and 4 of beef flavour.
Sam takes a packet out of the box without looking.
What is the probability that he picks a packet of chicken-flavoured food?

There are 20 packets in a box of cat food.

*   Using words:
    Only 2 of the 20 packets are chicken.
    So, picking chicken is unlikely.
*   Using a fraction:
    Two of the 20 packets are chicken.
    The probability of picking chicken is $\frac{2}{20}$, or $\frac{1}{10}$.
*   Using a ratio:
    The probability of picking chicken is $\frac{1}{10}$.
    We can write this as the part-to-whole ratio 1:10.
*   Using a percent:
    To express $\frac{2}{20}$ as a percent, find an equivalent
    fraction with denominator 100.

$$\frac{2}{20} \overset{\times 5}{\underset{\times 5}{=}} \frac{10}{100}, \text{ or } 10\%$$

> Recall that you can also use a calculator to help you write a fraction as a percent.

    The chance of picking chicken is 10%.

> When we express a probability as a percent, we often use the word *chance* to describe it.

When *all* the outcomes are favourable to an event, then the fraction:

Number of outcomes favourable to that event
Number of possible outcomes

has numerator equal to denominator, and the probability is 1, or 100%.
For example, the probability of picking a packet of cat food is: $\frac{20}{20} = 1$

When *no* outcomes are favourable to an event, then the fraction:

Number of outcomes favourable to that event
Number of possible outcomes

has numerator equal to 0, and the probability is 0, or 0%.
For example, the probability of picking
a packet of pork-flavoured cat food is: $\frac{0}{20} = 0$

The probability that an **impossible event** will occur is 0, or 0%.
The probability that a **certain event** will occur is 1, or 100%.
All other probabilities lie between 0 and 1.

Example

Twenty-five cans of soup were immersed in water.
Their labels came off so the cans now look identical.
There are: 2 cans of chicken soup; 4 cans of celery soup;
5 cans of vegetable soup; 6 cans of mushroom soup;
and 8 cans of tomato soup.
One can is picked, then opened.

a) What is the probability of each event?
   Write each probability as a ratio, fraction, and percent.
   i) The can contains celery soup.
   ii) The can contains fish.
   iii) The can contains celery soup or chicken soup.
   iv) The can contains soup.

b) State which event in part a is:
   i) certain   ii) impossible

A Solution

a) There are 25 cans, so there are 25 possible outcomes.
   i) Four cans contain celery soup.
      The probability of opening a can of celery soup is:
      4:25, or $\frac{4}{25} = \frac{16}{100}$, or 16%
   ii) None of the cans contain fish.
      The probability of opening a can of fish is: 0, or 0%
   iii) Four cans contain celery soup and two contain chicken soup.
      This is 6 cans in all.
      The probability of opening a can of celery soup or chicken soup is:
      6:25, or $\frac{6}{25} = \frac{24}{100}$, or 24%
   iv) Since all the cans contain soup, the probability of opening a can of soup is:
      20:20, or $\frac{20}{20}$, or 100%

b) i) The event that is certain to occur is opening a can that contains soup.
      This event has the greatest probability, 1.
   ii) The event that is impossible is opening a can that contains fish.
      This event has the least probability, 0.

Use a calculator when you need to.

1. Write the probability of each event as many different ways as you can.

   a) Roll a 3 or 5 on a die labelled 1 to 6.

   b) January immediately follows June.

   c) Pick an orange out of a basket that contains 2 oranges, 6 apples, and 8 peaches.

   d) The sun will set tomorrow.

2. A bag contains these granola bars: 12 apple, 14 peanut butter, 18 raisin, and 10 oatmeal. You pick one bar at random. Find the probability of picking:

   a) a peanut butter granola bar

   b) an apple granola bar

3. Two hundred fifty tickets for a draw were sold. One ticket, drawn at random, wins the prize.

   a) Joe purchased 1 ticket.
   What is the probability Joe will win?

   b) Maria purchased 10 tickets.
   What is the probability Maria will win?

   c) Ivan purchased 25 tickets.
   What is the probability Ivan will *not* win?

   Express each probability three ways.

4. Thanh has 20 felt pens in a pencil case.
He has 6 blue pens, 5 red pens, 2 yellow pens, 3 green pens, 2 brown pens, 1 purple pen, and 1 orange pen.
Thanh reaches into the case without looking and pulls out one pen.
Write a ratio, fraction, and percent to describe the probability that Thanh picks:

   a) either a yellow or a green pen

   b) either a blue or a red pen

   c) a coloured pen

   d) a grey pen

   e) a purple pen

5. The names of 8 students are in a hat.
You pick one name without looking.
Find each probability.
Express each probability as many ways as you can.

a) Laura will be picked.
b) Jorge will *not* be picked.
c) A three-letter name will be picked.
d) A five-letter name will be picked.
e) A name with 4 or more vowels will be picked.
f) A boy's or a girl's name will be picked.

6. Think of an experiment for which an event occurs with each probability.
Explain your choice.

a) 100%   b) $\frac{1}{2}$ c) 1:6   d) 0

7. **Assessment Focus**   Construct a spinner with red, yellow, blue, and
green sectors, so the following probabilities are true.
• The probability of landing on red is $\frac{1}{5}$.
• The probability of landing on yellow is 50%.
• The probability of landing on blue is 1:10.
• The probability of landing on green is $\frac{2}{10}$.
Explain how you drew your spinner.

8. **Take It Further**   A box contains 3 red, 2 green, and 4 white candies.
Carmen picked one candy, found it was white, and ate it.
She picked a second candy at random, found it was red, and ate it.
Carmen picked a third candy at random.

a) Which colour is the third candy most likely to be? Explain.
b) Write the probability that the third candy will be the colour named in
part a. Use a ratio, fraction, and percent to write the probability.
c) What is the probability that the candy will *not* be the colour named
in part a?

Reflect

The weather forecast shows a 90% chance of rain tomorrow.
How would this affect your plans for a class picnic? Why?

7.6 Tree Diagrams

Recall that an outcome is the possible result of an experiment or action.
When you roll a die, the outcomes are equally likely.
When you toss a coin, the outcomes are equally likely.
Some experiments have two or more actions.

Explore

You will need a die labelled 1 to 6 and a coin.

➤ List the possible outcomes of rolling the die and
tossing the coin.
How many possible outcomes are there?
How many outcomes include rolling a 4? Tossing a head?
➤ What is the theoretical probability of the event
"a head on the coin and a 2 on the die?"
➤ Conduct the experiment.
One of you tosses the coin and one rolls the die.
Record the results.
Calculate the experimental probability of the event
"a head on the coin and a 2 on the die" after each
number of trials.
- 10 trials
- 20 trials
- 50 trials
- 100 trials

➤ How do the experimental and theoretical probabilities compare?

Reflect & Share

Compare the strategy you used to find the outcomes with that of another pair
of classmates. Was one strategy more efficient than another? Explain.
Compare your probabilities. Combine your results to get 200 trials.
What is the experimental probability of the event
"a head on the coin and a 2 on the die?"
How do the experimental and theoretical probabilities compare?

Two events are **independent events** if the result of the one event
does not depend on the result of the other event.
Tossing two coins is an example of two independent events.
The outcome of the first toss does not affect the outcome of the second toss.
The outcome of the second toss does not depend on the outcome of the first toss.
We can use a **tree diagram** to show the possible outcomes for an experiment
that has two independent events.

When 2 coins are tossed, the outcomes for each coin are heads (H) or tails (T).
List the outcomes of the first coin toss.
This is the first branch of the tree diagram.
For each outcome, list the outcomes of the second coin toss.
This is the second branch of the tree diagram.
Then list the outcomes for the coins tossed together.

You could also use a table
to list the outcomes.

1st Coin	2nd Coin	Outcomes
H	H	HH
	T	HT
T	H	TH
	T	TT

	First Coin	
	H	**T**
H	HH	HT
T	TH	TT

Second Coin

There are 4 possible outcomes: HH, HT, TH, TT

Example

On this spinner, the pointer is spun once.
The colour is recorded.
The pointer is spun a second time.
The colour is recorded.

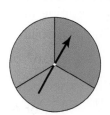

a) Draw a tree diagram to list the possible outcomes.
b) Find the probability of getting the same colours.
c) Find the probability of getting different colours.
d) Carina and Paolo carry out the experiment 100 times.
   There were 41 same colours and 59 different colours.
   How do the experimental probabilities compare to the
   theoretical probabilities? Explain.

A Solution

a) The first branch of the tree diagram lists the equally likely outcomes
of the first spin: blue, green, pink
The second branch lists the equally likely outcomes
of the second spin: blue, green, pink
For each outcome from the first spin,
there are 3 possible outcomes for the second spin.
Follow the paths from left to right. List all the possible outcomes.

First Spin	Second Spin	Possible Outcomes
blue	blue	blue/blue
	green	blue/green
	pink	blue/pink
green	blue	green/blue
	green	green/green
	pink	green/pink
pink	blue	pink/blue
	green	pink/green
	pink	pink/pink

Start

A list of all outcomes is the *sample space.*

b) From the tree diagram, there are 9 possible outcomes.
3 outcomes have the same colours: blue/blue, green/green, pink/pink
The probability of the same colour is:
$\frac{3}{9} = \frac{1}{3} \doteq 0.33$, or about 33%

c) 6 outcomes have different colours: blue/green, blue/pink,
green/blue, green/pink, pink/blue, pink/green
The probability of different colours is:
$\frac{6}{9} = \frac{2}{3} \doteq 0.67$, or about 67%

d) The experimental probability of the same colour is:
$\frac{41}{100} = 0.41$, or 41%
The experimental probability of different colours is:
$\frac{59}{100} = 0.59$, or 59%
These probabilities are different from the theoretical probabilities.
The greater the number of times the experiment is carried out,
the closer the theoretical and experimental probabilities may be.

1. List the sample space for each pair of independent events.
Why are the events independent?
a) Rolling a die labelled 3 to 8 and tossing a coin

b) Rolling a tetrahedron labelled 1 to 4 and spinning a pointer on this spinner

c) Rolling a pair of dice labelled 1 to 6

2. Use the outcomes from question 1a. Aseea wins if an odd number or a head shows.
Roberto wins if a number less than 5 shows. Who is more likely to win? Explain.

3. Use the outcomes from question 1b.
Name an outcome that occurs about one-half of the time.

4. Use the outcomes from question 1c.
How often are both numbers rolled greater than 4?

5. Hyo Jin is buying a new mountain bike.
She can choose from 5 paint colours—black, blue, red, silver, or gold—
and 2 seat colours—grey or black.
a) Use a table to display all the possible combinations of paint and seat colours.
b) Suppose Hyo Jin were to choose colours by pointing at lists without looking.
What is the probability she would end up with a silver
or black bike with a grey seat?

6. (Assessment Focus) Tara designs
the game *Mean Green Machine*.
A regular tetrahedron has its 4 faces
coloured red, pink, blue, and yellow.
A spinner has the colours shown.
When the tetrahedron is rolled,
the colour on its face down is recorded.
A player can choose to:
- roll the tetrahedron and spin the pointer, or
- roll the tetrahedron twice, or
- spin the pointer twice

To win, a player must make green by getting blue and yellow.
With which strategy is the player most likely to win?
Justify your answer. Play the game to check.
Show your work.

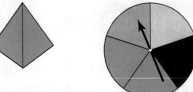

7. An experiment is: rolling a regular octahedron, labelled 1 to 8,
and drawing a counter from a bag that contains 4 counters:
green, red, yellow, blue
The number on the octahedron and the colour of the counter are recorded.

a) Use a regular octahedron, and a bag that contains the counters listed above.
Carry out the experiment 10 times. Record your results.

b) Combine your results with those of 9 classmates.

c) What is the experimental probability of each event?
   i) green and a 4
   ii) green or red and a 7
   iii) green or yellow and an odd number

d) Draw a tree diagram to list the possible outcomes.

e) What is the theoretical probability of each event in part c?

f) Compare the theoretical and experimental probabilities
of the events in part c.
What do you think might happen if you carried out
this experiment 1000 times? Explain.

Reflect

Which method do you prefer to find the sample space?
Why?

Math Link

Your World
The Canadian Cancer
Society runs a lottery every
year to raise money for
cancer research in Canada.
One year, the chances of
winning were given as the
ratio 1:12. This could also
be represented by the
fraction $\frac{1}{12}$.

All the Sticks

This game is based on a game originally played by the Blackfoot Nation.
The original materials were 4 animal bones and sticks.

HOW TO PLAY THE GAME:

1. Decorate:
   - 2 popsicle sticks with a zigzag pattern on one side
   - 1 popsicle stick with a circle pattern on one side
   - 1 popsicle stick with a pattern of triangles on one side
   Leave the other side of each popsicle stick blank.
2. Decide who will go first.
3. Place the counters in a pile on the floor.
4. Hold the 4 popsicle sticks in one hand,
   then drop them to the floor.
   Points are awarded according to the patterns
   that land face up.
   Find your pattern in the chart to determine your score.
   Take that number of counters.
   For example, if you score 4 points, take 4 counters.
   Take the counters from the pile until it has gone,
   then take counters from each other.

YOU WILL NEED

4 popsicle sticks (or tongue depressors); markers;
12 counters

NUMBER OF PLAYERS

2

GOAL OF THE GAME

To get all 12 counters

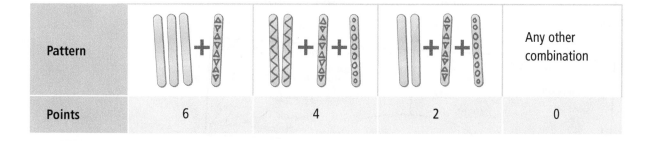

Pattern	(6 points image)	(4 points image)	(2 points image)	Any other combination
Points	6	4	2	0

What is the theoretical probability of scoring 6 points? How many points are you most likely to score in one turn? How did you find out?

5. Take turns.
   The first player to have all 12 counters wins.

Using a Frayer Model

Every new math topic has new words and ideas.

You can use a **Frayer Model** to help you remember new words and to better understand new ideas. A Frayer Model helps you make connections to what you already know.

Here is an example of a Frayer Model.

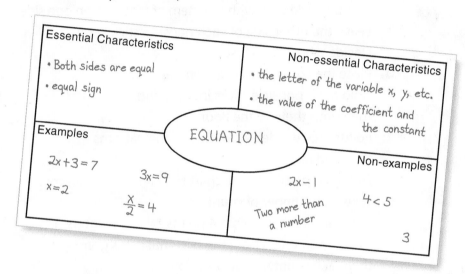

This model lists the essential and non-essential characteristics of the word.

What word do you think belongs in the centre of this Frayer Model? Explain.

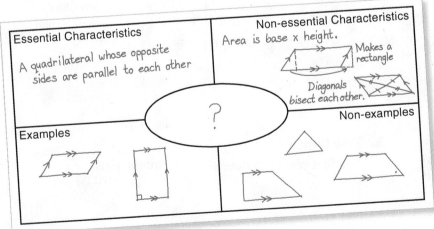

✓ Check

Work with a partner.

1. With your partner, make a Frayer Model
for 3 of these words.

- Mean
- Median
- Mode
- Range
- Outliers
- Independent events
- Probability
- Tree diagram

Choose the Frayer Model type
that best suits each word.

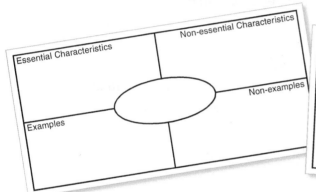

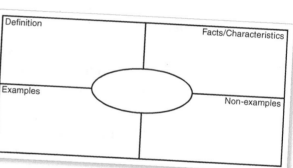

**This model gives the definition of the
word, then lists its characteristics.**

2. Work on your own.
Choose a word you did not use in question 1.
Each of you should choose a different word.
Make a Frayer Model.

3. Share your Frayer Model with your partner.
- Suggest ways your partner could improve his Frayer Model.
- Make changes to reflect your partner's suggestions.

4. Discuss with your partner.
- Did making Frayer Models help you understand
the new words and ideas? Explain.
- Do you think your Frayer Models will be a useful
review tool? Explain.

Unit Review

What Do I Need to Know?

- ☑ Each of the *mean*, *median*, and *mode* is a measure of central tendency, or an average.

- ☑ In a set of data:
  The *mean* is the sum of the numbers divided by the number of numbers in the set.
  The *median* is the middle number when the data are arranged in order. When there is an odd number of data in the set, the median is the middle number. When there is an even number of data, the median is the mean of the two middle numbers.
  The *mode* is the number that occurs most often.
  A set of data can have no mode, one mode, or more than one mode.
  The *range* is the difference between the greatest and least numbers.
  An *outlier* is a number that is much greater than or much less than most of the numbers in the data set.

- ☑ Probability Range:
  The probability of an event that is impossible is 0, or 0%.
  The probability of an event that is certain is 1, or 100%.
  All probabilities lie within this range.

- ☑ Two events are independent if the result of one event does not depend on the result of the other event.

What Should I Be Able to Do?

LESSON

7.1
7.2

1. To celebrate his birthday, Justin and his friends played miniature golf. Here are their scores:
29, 33, 37, 24, 41, 38, 48, 26, 36, 33, 40, 29, 36, 22, 31, 38, 42, 35, 33
It was a par 36 course.
This means that a good golfer takes 36 strokes to complete the course.

a) How many scores were under par? At par? Over par?
b) Find the range of the scores.
c) Calculate the mean, median, and mode scores.

7.2 **2.** The median shoe size of eight 12-year-old boys is $6\frac{1}{2}$. What might the shoe sizes be? Explain your answer.

7.3 **3.** Josephine recorded the hours she worked each week at her part-time job, for 10 weeks.
Here are the hours:
15, 12, 16, 10, 15, 15, 3, 18, 12, 10
a) Calculate the mean, median, and mode hours.
b) Identify the outlier. How might you explain this value?
c) Calculate the mean, median, and mode hours without the outlier. How is each measure affected when the outlier is not included?
d) Should the outlier be used in reporting the average number of hours Josephine worked? Explain.

7.4 **4.** The times, in minutes, that 14 students spent doing math homework over the weekend are:
27, 36, 48, 35, 8, 40, 41, 39, 74, 47, 44, 125, 37, 47
a) Calculate the mean, median, and mode for the data.
b) What are the outliers? Justify your choice. Calculate the mean without the outliers. What do you notice? Explain.
c) Which average best describes the data? Explain.
d) Should the outliers be used in reporting the average? Explain.

5. Annette's practice times for a downhill ski run, in seconds, are:
122, 137, 118, 119, 124, 118, 120, 118
a) Find the mean, median, and mode times.
b) Which measure best represents the times? Explain.
c) What is the range?
d) What time must Annette get in her next run so the median is 120 s? Explain.
e) What time must Annette get in her next run so the mean is 121 s? Is this possible? Explain.

6. Which average best describes a typical item in each data set? Explain.
a) size of sandals sold in a shoe store
b) test scores of a Grade 7 class
c) salaries in a company with 1000 employees
d) salaries in a company with 6 employees

7.5 **7.** Twenty cards are numbered from 1 to 20. The cards are shuffled. A card is drawn. Find the probability that the card has:
a) an odd number
b) a multiple of 4
c) a prime number
d) a number greater than 30
e) a number divisible by 1
Express each probability as many ways as you can.

8. In a board game, players take turns to spin pointers on these spinners. The numbers the pointers land on are multiplied. The player moves that number of squares on the board.

a) List the possible products.

b) What is the probability of each product in part a?

c) Which products are equally likely? Explain.

d) What is the probability of a product that is less than 10? Explain.

9. A spinner has 3 equal sectors labelled D, E, and F. A bag contains 3 congruent cubes: 1 green, 1 red, and 1 blue

The pointer is spun and a cube is picked at random.

a) Use a tree diagram to list the possible outcomes.

b) What is the probability of:
   i) spinning E?
   ii) picking a green cube?
   iii) spinning E and picking a green cube?
   iv) spinning D and picking a red cube?

10. Conduct the experiment in question 9.

a) Record the results for 10 trials.
   i) State the experimental probability for each event in question 9, part b.
   ii) How do the experimental and theoretical probabilities compare?

b) Combine your results with those of 9 other students. You now have the results of 100 trials. Repeat part a.

c) What happens to the experimental and theoretical probabilities of an event when the experiment is repeated hundreds of times?

11. The student council runs a coin toss game during School Spirit week. All the profits go to charity. Each player pays 50¢ to play. A player tosses two coins and wins a prize if the coins match. Each prize is $2.00. Can the student council expect to make a profit? Justify your answer.

Practice Test

1. The data show the time, in seconds, for swimmers to swim a 400-m freestyle race.

   208, 176, 265, 222, 333, 237, 225, 269, 303, 295, 238, 175, 257, 208, 271, 210

   a) What is the mean time?
   b) What is the mode time?
   c) What is the range of the times?
   d) What is the median time?

2. A sports store carries women's ice skates, sizes 4 to 10.
   In a particular 4-h period, these skate sizes are sold:

   4, 9, 8.5, 7.5, 6, 7, 6.5, 7, 7.5, 9, 8, 8, 18, 6.5, 8.5, 7, 5, 7, 9.5, 7

   a) Calculate the mean, median, and mode sizes.
      Which average best represents the data? Explain.
   b) Identify the outlier. How can you explain this size?
   c) Calculate the mean, median, and mode sizes without the outlier.
      How is each average affected when the outlier is not included?
   d) Should the outlier be used when the sales clerk reports the
      average skate size sold? Explain.

3. Match each probability to an event listed below.

   a) 0   b) $\frac{1}{2}$ c) 100%   d) 1:4
      i) You roll an even number on a die labelled 1 to 6.
      ii) You pick an orange out of a bag of apples.
      iii) You draw a red counter from a bag that contains
           3 yellow counters and 1 red counter.
      iv) You roll a number less than 7 on a die labelled 1 to 6.

4. Two dice are rolled. Each die is labelled 1 to 6.
   The lesser number is subtracted from the greater number,
   to get the difference. For example: $5 - 3 = 2$
   a) List the possible differences.
   b) Express the probability of each difference as a ratio, fraction, and percent.
      i) exactly 1   ii) greater than 3   iii) an odd number
   c) Carry out this experiment 10 times. Record your results.
   d) What is the experimental probability of each event in part b?
   e) Compare the theoretical and experimental probabilities.
      What do you notice?

Board Games

Many board games involve rolling pairs of dice, labelled 1 to 6.
Suppose you are on a particular square of a game board.
You roll the dice. You add the numbers on the uppermost faces.
How likely are you to roll 7?
Investigate to find out.

Part 1

Work with a partner.
Copy and complete this table.
Show the sums when two dice are rolled.

Sum of Numbers on Two Dice						
Number on Die	1	2	3	4	5	6
1	2	3	4	5	6	7
2				6		
3						
4	5					
5						
6						

When one die shows 1 and the other die shows 4, then the sum is 5.

How many different outcomes are there?
In how many ways can the sum be 6? 9? 2? 12?
Why do you think a table was used instead of a tree diagram?
Find the theoretical probability for each outcome.

Part 2

Work in groups of four.
1. Choose one of the following options:
    • Each of you rolls a pair of dice 25 times.
      Combine the results of your group.
    • Use a website your teacher gives you. Use the website to
      simulate rolling a pair of dice 100 times.
    Record the sum each time.

    Calculate the experimental probability of each outcome
    after 100 rolls.

2. Repeat *Step 1* three more times, to get results for 200 rolls, 300 rolls, and 400 rolls.

Calculate the experimental probability of each outcome after each number of rolls.

Summarize your results for 400 rolls in a table.

Sum	Number of Times Sum Occurred	Experimental Probability	Theoretical Probability
2			
3			
4			
5			

Find the mode of the sum rolled.
You can use a software program to help.
What does this tell you?

Check List

Your work should show:

✓ all records and calculations in detail

✓ clear explanations of your results, with the correct use of language of data analysis

✓ tables to show outcomes and probabilities

✓ an explanation of who is more likely to win

Part 3

What happens to the experimental probabilities as the number of rolls increases?

How does the experimental probability of each outcome compare with the theoretical probability? Explain.

Part 4

To win a board game, you must land on "home."
Suppose you are 7 squares from "home."
Your opponent is 4 squares from "home."
Who is more likely to win on the next roll? Explain.
Why do you think 7 is a lucky number?

Reflect on Your Learning

Why do you think you are learning about data analysis?

Many artists use geometric concepts in their work.

Think about what you have learned in geometry. How do these examples of First Nations art and architecture show geometry ideas?

What You'll Learn

- Identify and construct parallel and perpendicular line segments.
- Construct perpendicular bisectors and angle bisectors, and verify the constructions.
- Identify and plot points in the four quadrants of a grid.
- Graph and describe transformations of a shape on a grid.

Why It's Important

- A knowledge of the geometry of lines and angles is required in art and sports, and in careers such as carpentry, plumbing, welding, engineering, interior design, and architecture.

Key Words

- parallel lines
- perpendicular lines
- line segment
- bisect
- bisector
- perpendicular bisector
- angle bisector
- coordinate grid
- Cartesian plane
- *x*-axis
- *y*-axis
- origin
- quadrants

8.1 Parallel Lines

Identify parallel line
segments in these photos.
How could you check
they are parallel?

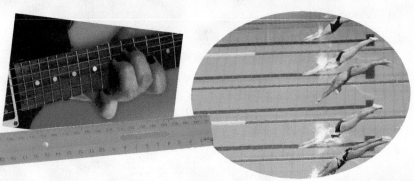

Explore

You may need a ruler, plastic triangle, tracing paper, protractor, and Mira.
Use any methods or tools. Draw a line segment on plain paper.
Draw a line segment parallel to the line segment.
Find as many ways to do this as you can using different tools.

Reflect & Share

Compare your methods with those of another pair of classmates.
How do you know the line segment you drew is parallel to the line segment?
Which method is most accurate? Explain your choice.

Connect

Parallel lines are lines on the same flat surface that never meet.
They are always the same distance apart.

Here are 3 strategies to draw a line segment parallel
to a given line segment.

➤ Use a ruler.
   Place one edge of a ruler along the line segment.
   Draw a line segment along
   the other edge of the ruler.

➤ Use a ruler and protractor.
   Choose a point on the line segment.
   Place the centre of the protractor on the point.

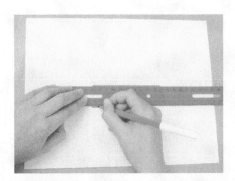

Align the base line of the protractor with
the line segment.
Mark a point at 90°.
Repeat this step once more.
Join the 2 points to draw a line segment parallel
to the line segment.

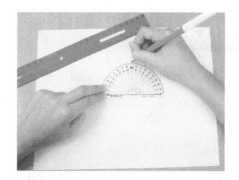

➤ Use a ruler and compass as shown below.

Example

Use a ruler and compass to draw a line segment parallel
to line segment AB that passes through point C.

A *line segment* is the
part of a line between
two points on the line.

A Solution

- Mark any point D on AB.

- Place the compass point on D.
  Set the compass so the pencil point is on C.
  Draw a circle.
  Label point E where the circle intersects AB.

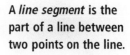

- Do not change the distance between
  the compass and pencil points.
  Place the compass point on E.
  Draw a circle through D.

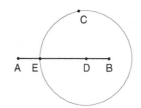

- Place the compass point on E.
  Set the compass so the pencil point is on C.

- Place the compass point on D.
  Draw a circle to intersect the circle
  through D.
  Label the point of intersection F.

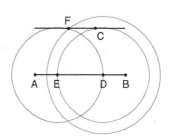

- Draw a line through points C and F.
  Line segment CF is parallel to AB.

The 2 line segments are parallel because they
are always the same distance apart.

Practice

1. Which lines are parallel? How do you know?

   a)   b)   c)

2. a) Draw line segment CD of length 5 cm.
      Use a ruler to draw a line segment parallel to CD.
   b) Choose 3 different points on CD.
      Measure the shortest distance from each point to the line segment you drew.
      What do you notice?

3. Draw line segment EF of length 8 cm.
   a) Use a ruler and protractor to draw a line segment parallel to EF.
   b) Use a ruler and compass to draw a line segment parallel to EF.

4. Suppose there are 2 line segments that look parallel.
   How could you tell if they are parallel?

5. Make a list of where you see parallel line segments
   in your community or around the house.
   Sketch diagrams to illustrate your list.

6. **Assessment Focus** Your teacher will give you
   a large copy of this diagram.
   Find as many pairs of parallel line segments as you can.
   How do you know they are parallel?

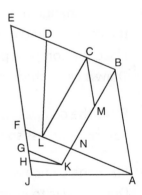

7. **Take It Further** Draw line segment CD.
   Use what you know about drawing parallel line segments
   to construct parallelogram CDEF.
   Explain how you can check you have drawn a parallelogram.

Reflect

Describe 3 different methods you can use to draw a line segment
parallel to a given line segment. Which method do you prefer?
Which method is most accurate? Explain your choice.

Perpendicular Lines

Identify perpendicular line segments in these photos. How could you check they are perpendicular?

Explore

You may need a ruler, plastic triangle, protractor, and Mira.
Use any methods or tools. Draw a line segment on plain paper.
Draw a line segment perpendicular to the line segment.
Find as many ways to do this as you can using different tools.

Reflect & Share

Compare your methods with those of another pair of classmates. How do you know the line segment you drew is perpendicular to the line segment?
Which method is most accurate? Explain your choice.

> **Recall that 2 lines intersect if they meet or cross.**

Connect

Two line segments are **perpendicular** if they intersect at right angles.
Here are 5 strategies to draw a line segment perpendicular to a given line segment.

➤ Use a plastic right triangle.
  Place the base of the triangle along the line segment.
  Draw a line segment along the side
  that is the height of the triangle.

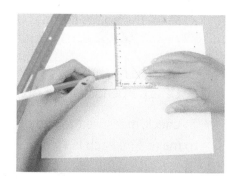

➤ Use paper folding.
  Fold the paper so that the line segment
  coincides with itself. Open the paper.
  The fold line is perpendicular to the line segment.

➤ Use a ruler and protractor.
Choose a point on the line segment.
Place the centre of the protractor on the point.
Align the base line of the protractor with
the line segment. Mark a point at 90°.
Join the 2 points to draw a line segment
perpendicular to the line segment.

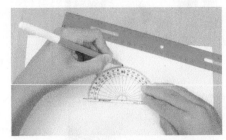

➤ Use a Mira. Place the Mira so that the reflection
of the line segment coincides with itself
when you look in the Mira.
Draw a line segment along the edge of the Mira.

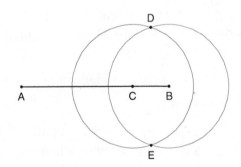

➤ Use a ruler and compass as shown below.

Example

Use a ruler and compass to draw a line segment
perpendicular to line segment AB.

A Solution

- Mark a point C on AB.

- Set the compass so the distance between
  the compass and pencil points is
  greater than one-half the length of CB.
  Place the compass point on B.
  Draw a circle that intersects AB.

- Do not change the distance between
  the compass and pencil points.
  Place the compass point on C.
  Draw a circle to intersect the first circle
  you drew.
  Label the points D and E where the
  circles intersect.

- Draw a line through points D and E.
  DE is perpendicular to AB.

To check, measure the angles
to make sure each is 90°.

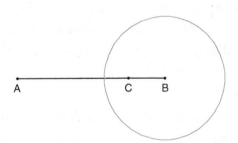

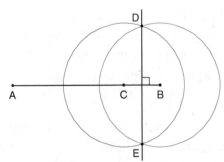

Practice

1. Which lines are perpendicular? How do you know?

   a)

   b)

   c)

2. a) Draw line segment AB of length 6 cm.
      Use a Mira to draw a line segment perpendicular to AB.
   b) Draw line segment CD of length 8 cm. Mark a point on the segment.
      Use paper folding to construct a line segment perpendicular to CD
      that passes through the point.
      How do you know that each line segment you drew is
      perpendicular to the line segment?

3. Draw line segment EF of length 10 cm.
   a) Use a ruler and protractor to draw a line segment perpendicular to EF.
   b) Use a ruler and compass to draw a line segment perpendicular to EF.
   c) Check that the line segments you drew are perpendicular to EF.

4. Make a list of where you see perpendicular line segments
   in the world around you. Sketch diagrams to illustrate your list.

5. **Assessment Focus**   Your teacher will give you
   a large copy of this diagram.
   Find as many pairs of perpendicular line segments as you can.
   How do you know they are perpendicular?

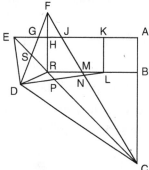

6. **Take It Further** Draw line segment JK of length 10 cm.
   Use what you know about drawing perpendicular and
   parallel line segments to construct a rectangle JKMN, where KM is 4 cm.
   Explain how you can check you have drawn a rectangle.

Reflect

Describe 4 different methods you can use to draw
a line perpendicular to a given line segment.
Which method do you prefer?
Which method is most accurate? Explain your choice.

Focus | Use a variety of methods to construct perpendicular bisectors of line segments.

Recall that a rhombus has all sides equal and opposite angles equal.

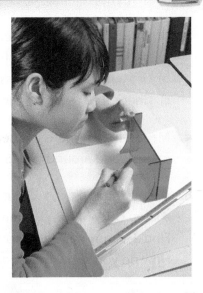

Each diagonal divides the rhombus
into 2 congruent isosceles triangles.
How do you know the triangles are isosceles?
How do you know the triangles are congruent?

You will investigate ways to cut line segments into 2 equal parts.

Explore

You may need rulers, protractors, tracing paper, plain paper,
and Miras.
Use any methods or tools.
Draw a line segment on plain paper.
Draw a line segment perpendicular to the line segment
that cuts the line segment in half.

Reflect & Share

Compare your results and methods with those of
another pair of classmates.
How could you use your method to cut your classmate's
line segment in half?

Connect

When you draw a line to divide a line segment into two
equal parts, you **bisect** the segment.
The line you drew is a **bisector** of the segment.

When the bisector is drawn at right angles to the segment,
the line is the **perpendicular bisector** of the segment.

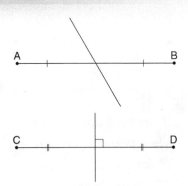

Here are 3 strategies to draw the perpendicular bisector of a given line segment.

➤ Use paper folding.
Fold the paper so that point A
lies on point B.
Crease along the fold. Open the paper.
The fold line is the perpendicular
bisector of AB.

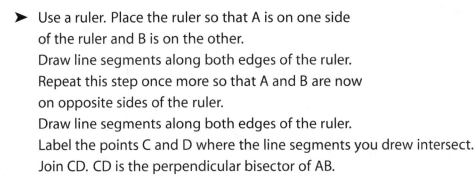

➤ Use a Mira. Place the Mira so that the reflection
of point A lies on point B.
Draw a line segment along the edge of the Mira.

➤ Use a ruler. Place the ruler so that A is on one side
of the ruler and B is on the other.
Draw line segments along both edges of the ruler.
Repeat this step once more so that A and B are now
on opposite sides of the ruler.
Draw line segments along both edges of the ruler.
Label the points C and D where the line segments you drew intersect.
Join CD. CD is the perpendicular bisector of AB.

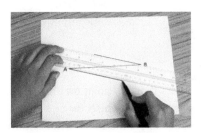 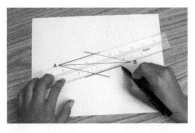 

Recall that each diagonal of a rhombus is a line of symmetry.
The diagonals intersect at right angles.
The diagonals bisect each other.
So, each diagonal is the perpendicular bisector of
the other diagonal.

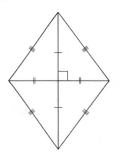

We can use these properties of a rhombus to construct
the perpendicular bisector of a line segment.
Think of the line segment as a diagonal of a rhombus.
As we construct the rhombus, we also construct the
perpendicular bisector of the segment.

Example

Use a ruler and compass to draw the perpendicular bisector of any line segment AB.

A Solution

Use a ruler and compass.

- Draw any line segment AB.

A ●————————● B

- Set the compass so the distance between the compass and pencil points is greater than one-half the length of AB.

- Place the compass point on A.
  Draw a circle.
  Do not change the distance between the compass and pencil points.
  Place the compass point on B.
  Draw a circle to intersect the first circle you drew.

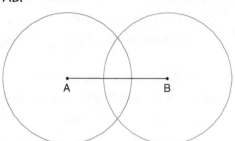

- Label the points C and D where the circles intersect.
  Join the points to form rhombus ACBD.
  Draw the diagonal CD.
  The diagonals intersect at E.
  CD is the perpendicular bisector of AB.
  That is, AE = EB and ∠AEC = ∠CEB = 90°

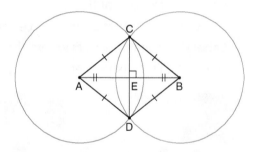

To check that the perpendicular bisector has been drawn correctly, measure the two parts of the segment to check they are equal, and measure the angles to check each is 90°.

> **Note that any point on the perpendicular bisector of a line segment is the same distance from the endpoints of the segment.**
> **For example, AC = BC and AD = BD**

Practice

Show all construction lines.

1. a) Draw line segment CD of length 8 cm.
   Use paper folding to draw its perpendicular bisector.

   b) Choose three different points on the bisector.
   Measure the distance to each point from C and from D.
   What do you notice?

2. a) Draw line segment EF of length 6 cm.
   Use a Mira to draw its perpendicular bisector.

   b) How do you know that you have drawn the perpendicular bisector of EF?

3. Draw line segment GH of length 4 cm.
   Use a ruler to draw its perpendicular bisector.

4. a) Draw line segment AB of length 5 cm.
   Use a ruler and compass to draw its perpendicular bisector.

   b) Choose three different points on the bisector.
   Measure the distance to each point from A and from B.
   What do you notice? Explain.

5. Find out what happens if you try to draw the perpendicular bisector of a line
   segment when the distance between the compass and pencil points is:
   a) equal to one-half the length of the segment
   b) less than one-half the length of the segment

6. Assessment Focus Draw line segment RS of length 7 cm.
   Use what you know about perpendicular bisectors to construct rhombus RTSU.
   How can you check that you have drawn a rhombus?

7. Look around you. Give examples of perpendicular bisectors.

8. Take It Further Draw a large △PQR.
   Construct the perpendicular bisector of each side.
   Label point C where the bisectors meet.
   Draw the circle with centre C and radius CP.

"Circum" is Latin for "around."
So, the *circumcircle* is the circle
that goes around a triangle.

9. Take It Further
   a) How could you use the construction in question 8 to draw
   a circle through any 3 points that do not lie on a line?
   b) Mark 3 points as far apart as possible. Draw a circle
   through the points. Describe your construction.

The point at which the
perpendicular bisectors
of the sides of a triangle
intersect is called the
circumcentre.

Reflect

How many bisectors can a line segment have?
How many perpendicular bisectors can a line segment have?
Draw a diagram to illustrate each answer.

Focus   Use a variety of methods to construct bisectors of angles.

You will investigate ways to divide an angle into 2 equal parts.

Explore

Your teacher will give you a large copy of this picture.
You may need rulers, protractors, tracing paper,
plain paper, and Miras.
Use any methods or tools.
George wants to share this slice of pie equally
with a friend.
Show how he could divide the slice of pie into
2 equal parts.

Reflect & Share

Compare your results and methods with those of
another pair of classmates.
How could you use your classmates' methods to divide
the slice of pie in half?

Connect

When you divide an angle into two equal parts, you *bisect* the angle.

Here are 3 strategies to draw the bisector of a given angle.

➤ Use paper folding.
  Fold the paper so that XY lies along ZY.
  Crease along the fold line.
  Open the paper.
  The fold line is the bisector of ∠XYZ.

➤ Use a Mira. Place the Mira so that the reflection
  of one arm of the angle lies along the other arm.
  Draw a line segment along the edge of the Mira.
  This line segment is the bisector of the angle.

➤ Use a plastic right triangle.

Place the triangle with one angle at B and one edge along BC.

Draw a line segment.

Place the triangle with the same angle at B and the same edge along AB.

Draw a line segment.

Label M where the line segments you drew intersect. BM is the bisector of ∠ABC.

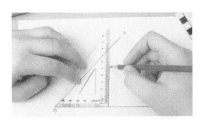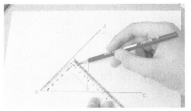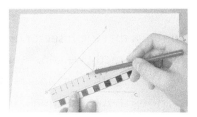

We can use the properties of a rhombus to construct the bisector of an angle.

Think of the angle as one angle of a rhombus.

Example

Draw obtuse ∠B of measure 126°.

Use a ruler and a compass to bisect the angle.

Measure the angles to check.

A Solution

Use a ruler and protractor to draw ∠B = 126°.

Use ∠B as one angle of a rhombus.

With compass point on B, draw a circle that
intersects one arm at F and the other arm at G.

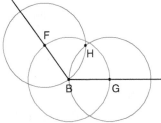

FB and BG are
2 sides of the
rhombus; FB = BG

Do not change the distance between
the compass and pencil points.

Place the compass point on F.

Draw a circle.

Place the compass point on G.

Draw a circle to intersect the second
circle you drew.

Label the point H where the circles intersect.

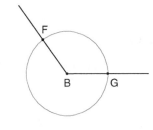

Join FH and HG to form rhombus BFHG.

Draw a line through BH.

This line is the **angle bisector** of ∠FBG.

That is, ∠FBH = ∠HBG

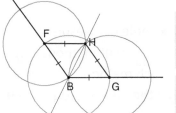

FH and HG are the
other 2 sides of
the rhombus.

BH is a diagonal
of the rhombus.

Use a protractor to check. Measure each angle.

∠FBG = 126°

∠FBH = 63° and ∠GBH = 63°

∠FBH + ∠GBH = 63° + 63°

= 126°

= ∠FBG

To check that the bisector of an angle has been drawn correctly, we can:

➤ Measure the two angles formed by the bisector.
They should be equal.

➤ Fold the angle so the bisector is the fold line.
The two arms should coincide.

➤ Place a Mira along the angle bisector.
The reflection image of one arm of the angle
should coincide with the other arm, and vice versa.

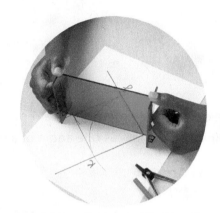

Practice

Show any construction lines.

1. Your teacher will give you a copy of this obtuse angle.
Use a Mira to bisect the angle.
Measure the two parts of the angle.
Are they equal?

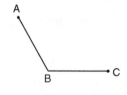

2. Your teacher will give you a copy of this acute angle.
Use a plastic right triangle to bisect the angle.
Measure the two parts of the angle.
Are they equal?

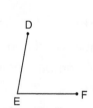

3. Use a ruler and compass.
   a) Draw acute ∠PQR = 50°. Bisect the angle.
   b) Draw obtuse ∠GEF = 130°. Bisect the angle.

4. Draw a reflex angle of measure 270°.
   a) How many different methods can you find to bisect this angle?
   b) Describe each method.
   Check that the bisector you draw using each method is correct.

A *reflex angle* is an angle between 180° and 360°.

5. You have used Miras, triangles, and paper folding to bisect an angle. What is the advantage of using a ruler and compass?

6. a) Draw line segment HJ of length 8 cm.
Draw the perpendicular bisector of HJ.
   b) Bisect each right angle in part a.
   c) How many angle bisectors did you need to draw in part b? Explain why you needed this many bisectors.

7. Assessment Focus Your teacher will give you a large copy of this isosceles triangle. Use a ruler and compass.

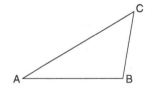

   a) Bisect ∠R.
   b) Show that the bisector in part a is the perpendicular bisector of ST.
   c) Is the result in part b true for:
   i) a different isosceles triangle?
   ii) an equilateral triangle?
   iii) a scalene triangle?
   How could you find out? Show your work.

8. Describe examples of angle bisectors that you see in the environment.

9. Take It Further Your teacher will give you a copy of this triangle.
Cut it out.
Fold the triangle so BC and BA coincide. Open the triangle.
Fold it so AB and AC coincide. Open the triangle.
Fold it so AC and BC coincide. Open the triangle.

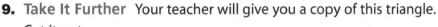

   a) Measure the angles each crease makes at each vertex. What do you notice?
   b) Label point K where the creases meet. Draw a circle in the triangle that touches each side of △ABC. What do you notice?
   c) What have you constructed by folding?

Reflect

How many bisectors can an angle have?
Draw a diagram to illustrate your answer.

Mid-Unit Review

8.1

1. a) Draw a line segment CD of length 9 cm. Draw a point F not on the line segment. Use a ruler and compass to construct a line segment parallel to CD that passes through F.

b) Draw line segment CD again. Use a different method to construct a line segment parallel to CD that passes through F.

c) Which method is more accurate? Explain you choice.

8.2

2. Your teacher will give you a large copy of this picture.

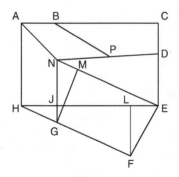

a) Identify as many parallel line segments as you can. How do you know they are parallel?

b) Find as many perpendicular line segments as you can. How do you know they are perpendicular?

8.3

3. a) Draw line segment AB of length 10 cm. Use a ruler and compass to draw the perpendicular bisector of AB.

b) Draw line segment AB again. Use a different method to draw the perpendicular bisector.

c) How can you check that you have drawn each bisector correctly?

4. a) Draw line segment AB of length 6 cm. AB is the base of a triangle.

b) Construct the perpendicular bisector of AB. Label point C where the perpendicular bisector intersects AB. Mark a point D on the perpendicular bisector. Join AD and DB.

c) What kind of triangle have you drawn? How do you know? What does CD represent?

8.4

5. a) Draw obtuse ∠PQR = 140°. Use a ruler and compass to bisect ∠PQR.

b) Draw acute ∠CDE = 50°. Use a different method to bisect ∠CDE.

c) How can you check that you have drawn each bisector correctly?

8.5   Graphing on a Coordinate Grid

Focus   Identify and plot points in four quadrants of a coordinate grid.

You have plotted points with whole-number coordinates on a grid.
Point A has coordinates (3, 2).
What are the coordinates of point B? Point C? Point D?

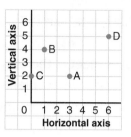

A vertical number line and a horizontal number line intersect at right angles at 0.
This produces a grid on which you can plot points with integer coordinates.

Explore

You will need grid paper and a ruler. Copy this grid.

➤ Plot these points: A(14, 0), B(6, 2), C(8, 8), D(2, 6), E(0, 14)
  Join the points in order.
  Draw a line segment from each point to the origin.
➤ Reflect the shape in the vertical axis.
  Draw its image.
  Write the coordinates of each vertex of the image.
➤ Reflect the original shape and the image in the horizontal axis.
  Draw the new image.
  Write the coordinates of each vertex of the new image.

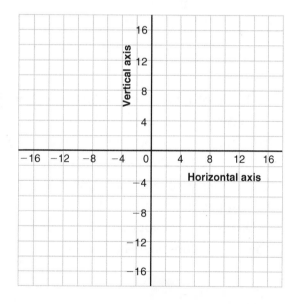

Your design should be symmetrical about the horizontal and vertical axes.
Describe the design. What shapes do you see?

Reflect & Share

Compare your design and its coordinates with those of another pair of classmates.
Describe any patterns you see in the coordinates of corresponding points.

A vertical number line and a horizontal number line that intersect
at right angles at 0 form a **coordinate grid**.
The horizontal axis is the **x-axis**.
The vertical axis is the **y-axis**.
The axes meet at the **origin**, (0, 0).
The axes divide the plane into four **quadrants**.
They are numbered counterclockwise.

This coordinate grid
is also called a
Cartesian plane.

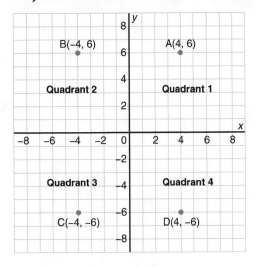

We do not need arrows
on the axes.

A pair of coordinates is
called an *ordered pair*.

In Quadrant 1, to plot point A, start at 4 on the x-axis and move up 6 units.
Point A has coordinates (4, 6).
In Quadrant 2, to plot point B, start at −4 on the x-axis and move up 6 units.
Point B has coordinates (−4, 6).
In Quadrant 3, to plot point C, start at −4 on the x-axis and move down 6 units.
Point C has coordinates (−4, −6).
In Quadrant 4, to plot point D, start at 4 on the x-axis and move down 6 units.
Point D has coordinates (4, −6).

We do not have
to include a
+ sign for a
positive
coordinate.

History
René Descartes lived in the 17th century.
He developed the coordinate grid.
It is named the Cartesian plane in his honour.
There is a story that René was lying in bed and watching a fly on the ceiling.
He invented coordinates as a way to describe the fly's position.

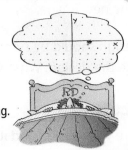

Example

a) Write the coordinates of each point.

    i) Q   **ii)** S

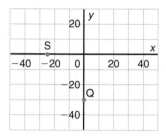

Notice that each grid square represents 10 units.

b) Plot each point on a grid.

    i) F(0, −15)   **ii)** G(−40, 0)

A Solution

a) Start at the origin each time.

    i) To get to Q, move 0 units right and 30 units down.
        So, the coordinates of Q are (0, −30).

    ii) To get to S, move 25 units left and 0 units down.
        So, the coordinates of S are (−25, 0).

Remember, first move left or right, then up or down.

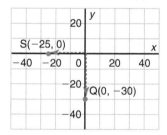

Point S is halfway between −20 and −30 on the *x*-axis.

b) **i)** F(0, −15)

       Since there is no movement left or right,
       point F lies on the *y*-axis.
       Start at the origin.
       Move 15 units down the *y*-axis. Mark point F.
       It is halfway between −10 and −20.

    ii) G(−40, 0)

       Start at −40 on the *x*-axis.
       Since there is no movement up or down,
       point G lies on the *x*-axis. Mark point G.

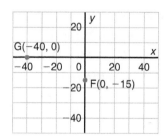

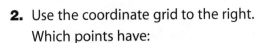

Practice

1. What is the scale on each axis?
   Write the coordinates of each point from A to K.

2. Use the coordinate grid to the right.
   Which points have:
   a) x-coordinate 0?
   b) y-coordinate 0?
   c) the same x-coordinate?
   d) the same y-coordinate?
   e) equal x- and y-coordinates?
   f) y-coordinate 2?

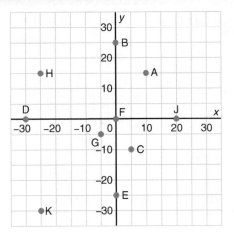

3. Draw a coordinate grid. Look at the ordered pairs below.
   Label the axes. How did you choose the scale?
   Plot each point.
   a) A(30, −30)   b) B(25, 0)   c) C(−10, 35)
   d) D(−15, 40)   e) E(15, 5)   f) F(0, −20)
   g) O(0, 0)   h) H(−20, −5)   i) I(−40, 0)
   Which point is the origin?

4. How could you use the grid in question 3 to plot these points?
   a) K(3, 5)   b) P(−10, 2)   c) R(−7, −8)

5. Which quadrant has all negative coordinates? All positive coordinates?
   Both positive and negative coordinates?

6. a) Plot these points: A(0, 5), B(−1, 4), C(−1, 3), D(−2, 3),
      E(−3, 2), F(−2, 1), G(−1, 1), H(−1, 0), J(0, −1), K(1, 0),
      L(1, 1), M(2, 1), N(3, 2), P(2, 3), R(1, 3), S(1, 4)
   b) Join the points in order. Then join S to A.
   c) Describe the shape you have drawn.

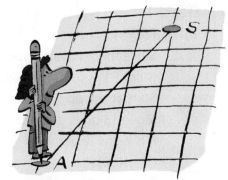

7. Draw a design on a coordinate grid.
   Each vertex should be at a point where grid lines meet.
   List the points used to make the design, in order.
   Trade lists with a classmate.
   Use the list to draw your classmate's design.

8. Use a 1-cm grid.
  a) Plot the points A(−3, 2) and B(5, 2).
   Join the points to form line segment AB.
   What is the horizontal distance between A and B?
   How did you find this distance?
  b) Plot the points C(3, −4) and D(3, 7).
   Join the points to form line segment CD.
   What is the vertical distance between C and D?
   How did you find this distance?

9. Use question 8 as a guide.
  Plot 2 points that lie on a horizontal or vertical line.
  Trade points with a classmate.
  Find the horizontal or vertical distance between
  your classmate's points.

10. Assessment Focus Use a coordinate grid.
  How many different parallelograms can you draw
  that have area 12 square units?
  For each parallelogram you draw, label its vertices.

11. a) Plot these points: K(−15, 20), L(5, 20), M(5, −10)
  b) Find the coordinates of point N that forms rectangle KLMN.

12. a) Plot these points on a grid: A(16, −14), B(−6, 12), and C(−18, −14).
   Join the points.
    What scale did you use? Explain your choice.
  b) Find the area of △ABC.

13. Take It Further The points A(−4, 4) and B(2, 4) are two vertices of a square.
  Plot these points on a coordinate grid.
  What are the coordinates of the other two vertices?
  Find as many different answers as you can.

Reflect

How did your knowledge of integers help you plot
points on a Cartesian plane?

Focus Graph translation and reflection images on a coordinate grid.

Recall that a translation slides a shape in a straight line. When the shape is on a square grid, the translation is described by movements right or left, and up or down.

> A translation and a reflection are transformations.

Which translation moved this shape to its image?

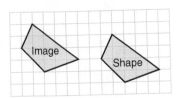

A shape can also be reflected, or flipped, in a mirror line. Where is the mirror line that relates this shape and its image?

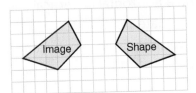

Explore

You will need 0.5-cm grid paper and a ruler.
Draw axes on the grid paper to get 4 quadrants.
Use the whole page. Label the axes.
Draw and label a quadrilateral.
Each vertex should be where the grid lines meet.

➤ Translate the quadrilateral. Draw and label the translation image. What do you notice about the quadrilateral and its image?

➤ Choose an axis. Reflect the quadrilateral in this axis. Draw and label the reflection image. What do you notice about the quadrilateral and its image?

➤ Trade your work with that of a classmate. Identify your classmate's translation. In which axis did your classmate reflect?

Reflect & Share

Did you correctly identify each transformation? Explain.
If not, work with your classmate to find the correct transformations.

Connect

➤ To translate △ABC 5 units right and 6 units down:
Begin at vertex A(−2, 5).
Move 5 units right and 6 units down to point A'(3, −1).
From vertex B(2, 3), move 5 units right and 6 units down
to point B'(7, −3).
From vertex C(−5, 1), move 5 units right and 6 units down
to point C'(0, −5).

We read A' as "A prime."

Each vertex of the
image is labelled with
a prime symbol.

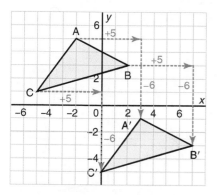

Then, △A'B'C' is the image of △ABC after a translation
5 units right and 6 units down.
△ABC and △A'B'C' are congruent.

➤ To reflect △ABC in the y-axis:
Reflect each vertex in turn.
The reflection image of A(−2, 5) is A'(2, 5).
The reflection image of B(2, 3) is B'(−2, 3).
The reflection image of C(−5, 1) is C'(5, 1).

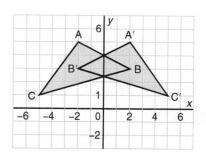

Then, △A'B'C' is the image of △ABC
after a reflection in the y-axis.
△ABC and △A'B'C' are congruent.
The triangles have different orientations:
we read △ABC clockwise; we read
△A'B'C' counterclockwise.

We could use a Mira to check the reflection.

Example

a) Plot these points: A(4, −4), B(6, 8), C(−3, 5), D(−6, −2)
   Join the points to draw quadrilateral ABCD.
   Reflect the quadrilateral in the *x*-axis.
   Draw and label the reflection image A′B′C′D′.

b) What do you notice about the line segment joining
   each point to its reflection image?

A Solution

a) To reflect quadrilateral ABCD in the *x*-axis:
   Reflect each vertex in turn.
   The reflection image of A(4, −4) is A′(4, 4).
   The reflection image of B(6, 8) is B′(6, −8).
   The reflection image of C(−3, 5) is C′(−3, −5).
   The reflection image of D(−6, −2) is D′(−6, 2).

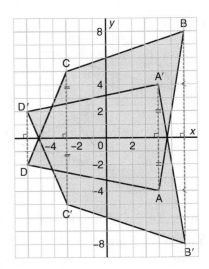

b) The line segments AA′, BB′, CC′, DD′ are vertical.
   The *x*-axis is the perpendicular bisector
   of each line segment.
   That is, the *x*-axis divides each line segment into
   2 equal parts, and the *x*-axis intersects each
   line segment at right angles.

In *Practice* question 7, you will investigate a similar reflection in the *y*-axis.

Practice

1. Identify each transformation. Explain your reasoning.

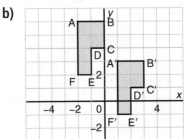

   a)   b)

2. Describe the horizontal and vertical distance required to move each point to its image.

a) A(5, −3) to A′(2, 6)   **b)** B(−3, 0) to B′(−5, −3)   **c)** C(2, −1) to C′(4, 3)

d) D(−1, 2) to D′(−4, 0)   **e)** E(3, 3) to E′(−3, 3)   **f)** F(4, −2) to F′(4, 2)

3. The diagram shows 4 parallelograms.

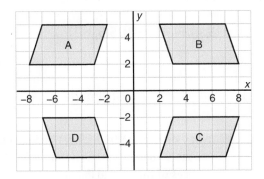

a) Are any 2 parallelograms related by a translation? If so, describe the translation.

b) Are any 2 parallelograms related by a reflection? If so, describe the reflection.

4. Copy this pentagon on grid paper.
Write the coordinates of each vertex.
After each transformation:

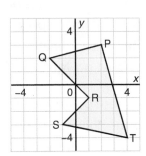

- Write the coordinates of the image of each vertex.
- Describe the positional change of the vertices of the pentagon.

a) Draw the image after a translation 3 units left and 2 units up.

b) Draw the image after a reflection in the x-axis.

c) Draw the image after a reflection in the y-axis.

5. Plot these points on a coordinate grid:
A(1, 3), B(3, −2), C(−2, 5), D(−1, −4), E(0, −3), F(−2, 0)

a) Reflect each point in the x-axis.
Write the coordinates of each point and its reflection image.
What patterns do you see in the coordinates?

b) Reflect each point in the y-axis.
Write the coordinates of each point and its reflection image.
What patterns do you see in the coordinates?

c) How could you use the patterns in parts a and b to check that you have drawn the reflection image of a shape correctly?

6. a) Plot the points in question 5.
Translate each point 4 units left and 2 units down.

b) Write the coordinates of each point and its translation image.
What patterns do you see in the coordinates?

c) How could you use these patterns to write the coordinates of
an image point after a translation, without plotting the points?

7. a) Plot these points on a coordinate grid: $P(1, 4), Q(-3, 4), R(-2, -3), S(5, -1)$
Join the points to draw quadrilateral PQRS.
Reflect the quadrilateral in the y-axis.

b) What do you notice about the line segment joining each point to its image?

8. Assessment Focus

a) Plot these points on a coordinate grid:
$A(2, 4), B(4, 4), C(4, 2), D(6, 2), E(6, 6)$
Join the points to draw polygon ABCDE.

> When there are 2 transformation images,
> we use a "double" prime notation for the
> vertices of the second image.

b) Translate the polygon 4 units right and 6 units up.
Write the coordinates of each vertex of the image polygon A′B′C′D′E′.

c) Reflect the image polygon A′B′C′D′E′ in the y-axis.
Write the coordinates of each vertex of the image polygon A″B″C″D″E″.

d) How does polygon A″B″C″D″E″ compare with polygon ABCDE?

9. a) Plot these points on a coordinate grid: $F(-5, 8), G(0, 8), H(-1, 5), J(-5, 5)$
Join the points to draw trapezoid FGHJ.

b) Translate the trapezoid 2 units right and 1 unit down.

c) Translate the image trapezoid F′G′H′J′ 2 units right and 1 unit down.

d) Repeat part c four more times for each resulting image trapezoid.

e) Describe the translation that moves trapezoid FGHJ to
the final image.

10. Take It Further Draw a shape and its image that could represent
a translation and a reflection. What attributes does the shape have?

Reflect

How is a translation different from a reflection?
How are these transformations alike?

Focus Graph rotation images on a coordinate grid.

Recall that a rotation turns a shape about a point of rotation.
The rotation may be clockwise or counterclockwise.
The point of rotation may be:

On the shape

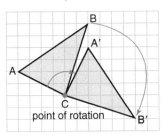

Off the shape

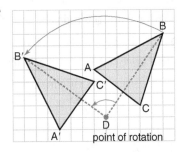

How would you describe each rotation?

Explore

You will need 0.5-cm grid paper, tracing paper, a protractor, and a ruler.
Draw axes on the grid paper to get 4 quadrants.
Place the origin at the centre of the paper.
Label the axes.
Plot these points: O(0, 0), B(5, 3), and C(5, 4)
Join the points in order, then join C to O.
Use the origin as the point of rotation.

➤ Rotate the shape 90° counterclockwise.
  Draw its image.

➤ Rotate the original shape 180° counterclockwise.
  Draw its image.

➤ Rotate the original shape 270° counterclockwise.
  Draw its image.

What do you notice about the shape and its 3 images?
What have you drawn?

Reflect & Share

Compare your work with that of another pair of classmates.
What strategies did you use to measure the rotation angle?
Would the images have been different if you had rotated clockwise
instead of counterclockwise? Explain your answer.

Connect

To rotate the shape at the right clockwise:

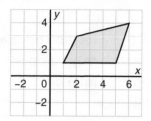

➤ Trace the shape and the axes.
  Label the positive y-axis on the tracing paper.
  Rotate the tracing paper clockwise about the origin
  until the positive y-axis coincides with the positive x-axis.
  With a sharp pencil, mark the vertices of the image.
  Join the vertices to draw the image after a 90° clockwise
  rotation about the origin, below left.

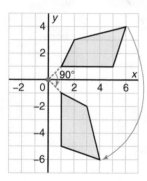

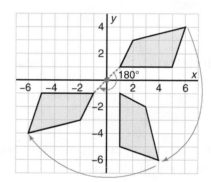

➤ Place the tracing paper so the shape coincides with its image.
  Rotate the tracing paper clockwise about the origin until
  the positive y-axis coincides with the negative y-axis.
  Mark the vertices of the image.
  Join the vertices to draw the image of the original shape after a
  180° clockwise rotation about the origin, above right.

➤ Place the tracing paper so the shape coincides
  with its second image.
  Rotate the tracing paper clockwise about the origin
  until the positive y-axis coincides
  with the negative x-axis.
  Mark, then join, the vertices of the image.
  This is the image after a 270° clockwise rotation
  about the origin.

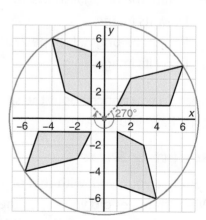

All 4 quadrilaterals are congruent.
A point and all its images lie on a circle, centre the origin.

Example

a) Plot these points: $B(-5, 6), C(-3, 4), D(-8, 2)$
   Join the points to draw $\triangle BCD$.
   Rotate $\triangle BCD$ 90° about the origin, O.
   Draw and label the rotation image $\triangle B'C'D'$.
b) Join C, D, C', D' to O.
   What do you notice about these line segments?

> A counterclockwise rotation is shown by a positive angle such as +90°, or 90°. A clockwise rotation is shown by a negative angle such as −90°.

A Solution

A rotation of 90° is a counterclockwise rotation.
a) Use tracing paper to draw the image $\triangle B'C'D'$.
   Rotate the paper counterclockwise until the
   positive y-axis coincides with the negative x-axis.
   After a rotation of 90° about the origin:
   $B(-5, 6) \rightarrow B'(-6, -5)$
   $C(-3, 4) \rightarrow C'(-4, -3)$
   $D(-8, 2) \rightarrow D'(-2, -8)$
b) From the diagram, $OC = OC'$ and $OD = OD'$
   $\angle COC' = \angle DOD' = 90°$

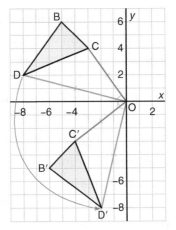

Practice

1. Each grid shows a shape and its rotation image.
   Identify the angle and direction of rotation, and the point of rotation.

   a)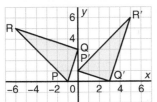

   b)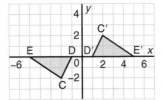

2. Identify the transformation that moves the shape in Quadrant 2 to each image. Explain how you know.

   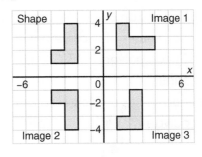

3. a) Copy △DEF on grid paper.

Write the coordinates of each vertex.

After each rotation:

- Write the coordinates of the image of each vertex.
- Describe the positional change of the vertices of the triangle.

b) Rotate △DEF −90° about the origin to its image △D'E'F'.

c) Rotate △DEF +270° about the origin to its image △D"E"F".

d) What do you notice about the images in parts b and c?
Do you think you would get a similar result with any shape that you rotate −90° and +270°? Explain.

4. Plot each point on a coordinate grid:

A(2, 5), B(−3, 4), C(4, −1)

a) Rotate each point 180° about the origin O to get image points A', B', C'.
Write the coordinates of each image point.

b) Draw and measure:
i) OA and OA'   **ii)** OB and OB'   **iii)** OC and OC'
What do you notice?

c) Measure each angle.
i) ∠AOA'   **ii)** ∠BOB'   **iii)** ∠COC'
What do you notice?

d) Describe another rotation of A, B, and C that would result in the image points A', B', C'.

5. Repeat question 4 for a rotation of −90° about the origin.

6. Assessment Focus

a) Plot these points on a coordinate grid:
A(6, 0), B(6, 2), C(5, 3), D(4, 2), E(2, 2), F(2, 0)
Join the points to draw polygon ABCDEF.

b) Translate the polygon 6 units left and 2 units up.
Write the coordinates of each vertex of the image polygon A'B'C'D'E'F'.

c) Rotate the image polygon A'B'C'D'E'F' 90° counterclockwise about the origin.
Write the coordinates of each vertex of the image polygon A"B"C"D"E"F".

d) How does polygon A"B"C"D"E"F" compare with polygon ABCDEF?

7. Draw a large quadrilateral in the 3rd quadrant.

   a) Rotate the quadrilateral 180° about the origin.

   b) Reflect the quadrilateral in the *x*-axis.
      Then reflect the image in the *y*-axis.

   c) What do you notice about the image in part a
      and the second image in part b?
      Do you think you would get a similar result if you started:
      i) with a different shape?   **ii)** in a different quadrant?
      Investigate to find out. Write about what you discover.

8. a) Plot these points on a coordinate grid:
      $R(-1, -1), S(-1, 4), T(2, 4), U(2, -1)$
      Join the points to draw rectangle RSTU.

   b) Choose a vertex to use as the point of rotation.
      Rotate the rectangle 90° counterclockwise.

   c) Repeat part b two more times for each image rectangle.

   d) Describe the pattern you see in the rectangles.

   e) Is there a transformation that moves rectangle RSTU to the final
      image directly? Explain.

9. Take It Further Plot these points: $C(2, 6), D(3, -3), E(5, -7)$

   a) Reflect △CDE in the *x*-axis to its image △C'D'E'.
      Rotate △C'D'E' $-90°$ about the origin to its image △C"D"E".

   b) Rotate △CDE $-90°$ about the
      origin to its image △PQR.
      Reflect △PQR in the *x*-axis
      to its image △P'Q'R'.

   c) Do the final images in
      parts a and b coincide?
      Explain your answer.

Reflect

When you see a shape and its transformation image on a grid,
how do you know what type of transformation it is?
Include examples in your explanation.

Using a Computer to Transform Shapes

Geometry software can be used to transform shapes.
Use available geometry software.

Open a new sketch. Check that the distance units
are centimetres. Display a coordinate grid.

Should you need help at any time, use the software's Help Menu.

Translating a Shape

Construct a quadrilateral ABCD.
Record the coordinates of each vertex.

Select the quadrilateral. Use the software to translate the
quadrilateral 3 units right and 2 units down. Record the
coordinates of each vertex of the translation image A'B'C'D'.

Rotating a Shape

Use the image quadrilateral A′B′C′D′.
Select a vertex of the quadrilateral as the point
of rotation. Select quadrilateral A′B′C′D′.
Rotate the quadrilateral 90° counterclockwise.
Record the coordinates of each vertex
of the rotation image A″B″C″D″.

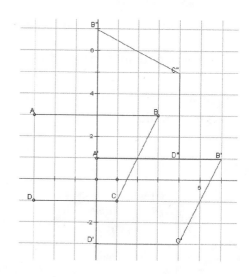

Reflecting a Shape

Use the image quadrilateral A″B″C″D″.
Select one side of the quadrilateral as the mirror line.
Select quadrilateral A″B″C″D″.
Reflect the quadrilateral in the mirror line.
Record the coordinates of each vertex of the reflection image A‴B‴C‴D‴.

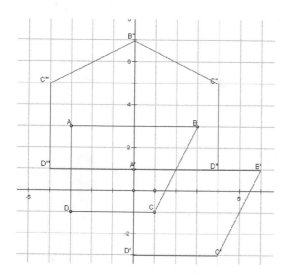

✓ Check

Create another shape.
Use any or all of the transformations above
to make a design that covers the screen.
Colour your design to make it attractive. Print your design.

Making a Study Card

There is a lot of important information in a math unit.

A study card can help you organize and review this information.

How to Start

Read through each lesson in this unit.
As you read, make a study sheet by writing down:
- the main ideas of each lesson
- any new words and what they mean
- any formulas
- one or two examples per lesson
- any notes that might help you remember the concepts

Making a Study Card

- Read over your study sheet.
- Decide which information to include on your study card.
  Only include information you need help remembering.
- Put the information on a recipe card.
  If you have too much information to fit on one card, use two or more cards.

Using Your Study Cards

- Use your study cards as you work through the Unit Review.
- When you have finished, you should have some information on your study cards that you do not need help remembering anymore.
- Make a final study card using only one recipe card.

Compare your study card with that of a classmate.
What information does your classmate have that you do not have?
Should everyone's study card be the same? Explain.

The more you use your study card, the less you are going to need it.

Here is an example of a study card for Geometry.

Your study card is your own.
Make it so it helps you.
It may be different from your classmate's study card.

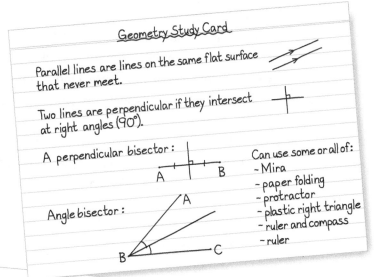

Geometry Study Card

Parallel lines are lines on the same flat surface that never meet.

Two lines are perpendicular if they intersect at right angles (90°).

A perpendicular bisector :

Angle bisector :

Can use some or all of:
- Mira
- paper folding
- protractor
- plastic right triangle
- ruler and compass
- ruler

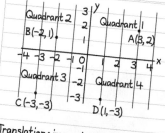

Translation : image is congruent and has same orientation.
Reflection : image is congruent and orientation is reversed.
Rotation : image is congruent and has same orientation.

Coordinates of point A(3, 2) after:
Translation of 4 units left and 1 unit up:
A(3, 2) $\longrightarrow$ A'(−1, 3)
Reflection in x−axis:
A(3, 2) $\longrightarrow$ A'(3, −2)
Reflection in y−axis:
A(3, 2) $\longrightarrow$ A'(−3, 2)
Rotation of 90° counterclockwise:
A(3, 2) $\longrightarrow$ A'(−2, 3)
Rotation of 90° clockwise:
A(3, 2) $\longrightarrow$ A'(2, −3)

Unit Review

☑ **Parallel lines**

are lines on the same flat surface that never meet.

☑ **Perpendicular lines**

Two lines are *perpendicular* if they intersect at right angles.

☑ **Perpendicular Bisector**

The *perpendicular bisector* of a line segment is drawn
at right angles to the segment and divides the
segment into two equal parts. Line segment CD
is the perpendicular bisector of segment AB.

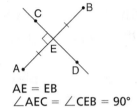

AE = EB
∠AEC = ∠CEB = 90°

☑ **Bisector of an Angle**

The *bisector of an angle* divides the angle into two equal
angles. Line segment QS is the bisector of ∠PQR.

∠PQS = ∠SQR

☑ **Transformations on a Coordinate Grid**

A point or shape can be:

- translated (slid)
- reflected in the *x*-axis or the *y*-axis (flipped)
- rotated about the origin (turned)

Math Links

Art
Origami is the Japanese name for the art of paper folding.
Use the library or the Internet to get instructions
to make a model. Fold a sheet of paper to make
your chosen model. Then, unfold the paper
and look at the creases. Label as many pairs
of parallel line segments as you can.
Do the same with perpendicular line segments,
perpendicular bisectors, and angle bisectors.

What Should I Be Able to Do?

LESSON

8.1 **1. a)** Draw line segment FG of length 5 cm.

   b) Mark a point H above FG. Draw a line segment parallel to FG that passes through point H.

   c) Mark a point J below FG. Draw a line segment parallel to FG that passes through point J.

   d) Explain how you can check that the line segments you drew in parts b and c are parallel.

8.2 **2. a)** Draw line segment CD of length 12 cm.

   b) Mark a point E above CD. Draw a line segment perpendicular to CD that passes through point E. Label point F where the line segment intersects CD.

   c) Join CE and ED. What does EF represent?

8.3 **3. a)** Draw line segment AB. Fold the paper to construct the perpendicular bisector.

   b) Draw line segment CD. Use a Mira to construct the perpendicular bisector.

   c) Draw line segment EF. Use a ruler and compass to construct the perpendicular bisector.

   d) Which of the three methods is most accurate? Justify your answer.

8.4 **4. a)** Draw acute $\angle BAC = 70°$. Fold the paper to construct the angle bisector.

   b) Draw right $\angle DEF$. Use a Mira to construct the angle bisector.

   c) Draw obtuse $\angle GHJ = 100°$. Use a ruler and compass to construct the angle bisector.

   d) Which method is most accurate? Justify your answer.

8.5 **5. a)** On a coordinate grid, plot each point. Join the points in order. Then join D to A. How did you choose the scale?
    $A(-20, -20)$ $B(30, -20)$
    $C(15, 30)$ $D(-35, 30)$

   b) Name the quadrant in which each point is located.

   c) Identify the shape. Find its area.

6. Do not plot the points. In which quadrant is each point located? How do you know?
   a) $A(6, -4)$ **b)** $B(-4, -2)$
   c) $C(-3, 2)$ **d)** $D(6, 4)$

7. a) Find the horizontal distance between each pair of points.
    i) $A(-5, 1)$ and $B(7, 1)$
    ii) $C(-2, -3)$ and $D(9, -3)$

   b) Find the vertical distance between each pair of points.
    i) $E(4, -5)$ and $F(4, 3)$
    ii) $G(-3, -6)$ and $H(-3, 0)$

8. Plot each point on a coordinate grid: A(−1, −1), C(3, 1)
A and C are opposite vertices of a rectangle. Find the coordinates of the other 2 vertices.

8.6 9. a) Plot these points on a coordinate grid.
P(3, 1), Q(7, 1), R(5, 3), S(3, 3)
Join the points to draw trapezoid PQRS.
How do you know it is a trapezoid? Explain.
b) Translate the trapezoid 4 units right. Write the coordinates of each vertex of the image trapezoid P′Q′R′S′.
c) Reflect trapezoid P′Q′R′S′ in the x-axis.
Write the coordinates of each vertex of the image trapezoid P″Q″R″S″.
d) How does trapezoid P″Q″R″S″ compare with trapezoid PQRS?

10. Repeat question 9. This time, apply the reflection before the translation. Is the final image the same? Explain.

8.7 11. a) Plot these points on a coordinate grid:
A (−2, 3)   B(−4, 0)
C(−2, −3)   D(2, −3)
Join the points to draw quadrilateral ABCD.

b) Draw the image of quadrilateral ABCD after each transformation:
   i) a translation 7 units left and 8 units up
   ii) a reflection in the x-axis
   iii) a rotation of +90° about the origin
c) How are the images alike? Different?

12. Use these points:
A(−2, 3), B(0, −1), and C(4, −3)
a) Suppose the order of the coordinates is reversed.
In which quadrant would each point be now?
Draw △ABC on a coordinate grid.
b) Which transformation changes the orientation of the triangle? Justify your answer with a diagram.
c) For which transformation is the image of AC perpendicular to AC? Justify your answer with a diagram.

13. a) Plot these points on a coordinate grid.
C(6, −3), D(−4, 3), E(6, 3)
Join the points to draw △CDE.
b) Translate △CDE 5 units left and 4 units up to image △C′D′E′.
c) Rotate △C′D′E′ +90° about the origin to image △C″D″E″.
d) How does △C″D″E″ compare with △CDE?

Practice Test

1. Look around the classroom. Where do you see:
   a) parallel line segments?   b) perpendicular line segments?
   c) perpendicular bisectors?   d) angle bisectors?

2. Your teacher will give you a copy
   of each of these triangles.
   a) Construct the bisector of each angle in △ABC.
      Use a different method or tool
      for each bisector.
   b) Construct the perpendicular bisector
      of each side of △DEF.
      Use a different method or tool
      for each bisector.
   Describe each method or tool used.

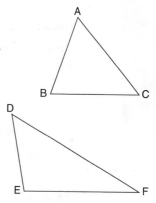

3. a) On a coordinate grid, draw a triangle with area 12 square units.
      Place each vertex in a different quadrant.
   b) Write the coordinates of each vertex.
   c) Explain how you know the area is 12 square units.
   d) Translate the triangle 6 units right and 3 units down.
      Write the coordinates of each vertex of the translation image.
   e) Reflect the triangle in the y-axis.
      Write the coordinates of each vertex of the reflection image.
   f) Rotate the triangle 90° clockwise about the origin.
      Write the coordinates of each vertex of the rotation image.

4. a) Plot these points on a coordinate grid.
      A(−2, 0), B(4, 0), C(3, 4), D(−1, 3)
      Join the points to draw quadrilateral ABCD.
   b) Translate quadrilateral ABCD 2 units left and 3 units down
      to the image quadrilateral A′B′C′D′.
   c) Translate quadrilateral A′B′C′D′ 6 units right and 7 units up
      to the image quadrilateral A″B″C″D″.
   d) Describe the translation that moves quadrilateral ABCD
      to quadrilateral A″B″C″D″.
   e) What would happen if the order of the translations was reversed?
      Explain your answer.

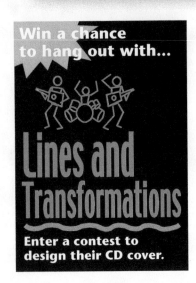

You have entered a contest to design the front and back covers of a CD for a new band called *Lines and Transformations*.

Part 1

Your design for the front cover will be created on 4 pieces of paper. It has to include:

- geometric shapes
- parallel line segments and perpendicular line segments
- geometric constructions

Work in a group of 4.

Brainstorm design ideas for the cover.

Sketch your cover. Show all construction lines.

Each person is responsible for one piece of the cover.

Make sure the pattern or design continues across a seam.

Draw your cover design.

Add colour to make it appealing.

Write about your design.

Explain your choice of design and how it relates to the geometric concepts of this unit.

Check List

Your work should show:

✓ a detailed sketch of the front cover, including construction lines

✓ a design for the back cover, using transformations

✓ your understanding of geometric language and ideas

✓ accurate descriptions of construction methods and transformations used

Part 2

Your design for the back cover will be created on a grid.
Each of you chooses a shape from your design for the front cover.
Draw the shape on a coordinate grid.
Use transformations to create a design with your shape.
Colour your design.

Write about your design.
Describe the transformations you used.
Record the coordinates of the original shape and 2 of the images.

Reflect on Your Learning

Write 3 things you now know about parallel and perpendicular line segments that you did not know before.
What have you learned about transformations?

Integer Probability

Work with a partner.

Four integer cards, labelled -3, -2, $+1$, and $+3$, are placed in a bag.

James draws three cards from the bag, one card at a time. He adds the integers.

Materials:
- four integer cards labelled -3, -2, $+1$, $+3$
- brown paper bag

James predicts that because the sum of all four integers is negative, it is more likely that the sum of any three cards drawn from the bag will be negative.

In this *Investigation*, you will conduct James' experiment to find out if his prediction is correct.

Part 1

➤ Place the integer cards in the bag.
  Draw three cards and add the integers.
  Is the sum negative or positive?
  Record the results in a table.

Integer 1	Integer 2	Integer 3	Sum

➤ Return the cards to the bag. Repeat the experiment until you have 20 sets of results.

➤ Look at the results
in your table.
Do the data support
James' prediction?
How can you tell?

➤ Combine your results
with those of 4 other
pairs of classmates.
You now have
100 sets of results.
Do the data support
James' prediction?
How can you tell?

➤ Use a diagram or other model to find the theoretical
probability of getting a negative sum.
Do the results match your experiment?

➤ Do you think the values of the integers make a difference?
Find 4 integers (2 positive, 2 negative) for which
James' prediction is correct.

Part 2

Look at the results of your investigation in *Part 1*.
➤ If the first card James draws is negative, does it affect
the probability of getting a negative sum?
Use the results of *Part 1* to support your thinking.

➤ If the first card James draws is positive, does it affect
the probability of getting a negative sum?
Use the results of *Part 1* to support your thinking.

UNIT

1

1. a) Use algebra.
    Write a relation
    for this Input/
    Output table.

Input *n*	Output
1	6
2	10
3	14
4	18

b) Graph the
    relation.

c) Describe the graph.

d) Explain how the graph
    illustrates the relation.

e) Suggest a real-life situation this
    graph could represent.

2. The Grade 7 students are organizing
    an end-of-the-year dance.
    The disc jockey charges a flat
    rate of $85. The cost to attend
    the dance is $2 per student.

a) How much will the dance cost
    if 30 students attend?
    50 students attend?

b) Write a relation for the cost of the
    dance when *s* students attend.

c) Suppose the cost of
    admission doubles.
    Write a relation for the total cost
    of the dance for *s* students.

d) Suppose the cost of the
    disc jockey doubles.
    Write a relation for the total cost
    of the dance for *s* students.

2

3. a) Write the addition equation
    modelled by each number line.

b) Describe a situation that each
    number line could represent.

i)

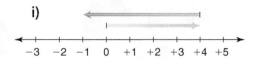

ii)

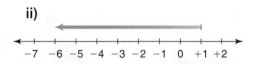

4. On January 11, the predicted high
    and low temperatures in Flin Flon,
    Manitoba were −4°C and −13°C.

a) Which is the high temperature and
    which is the low temperature?

b) What is the difference in
    temperatures?

3

5. Use front-end estimation to
    estimate each sum or difference.

a) 7.36 + 2.23   **b)** 4.255 − 1.386

c) 58.37 − 22.845 **d)** 217.53 + 32.47

6. A store has a sale.
    It will pay the tax if your
    purchase totals $25 or more.
    Justin buys a computer game for
    $14.95, some batteries for $7.99,
    and a gaming magazine for $5.95.

a) How much money did Justin
    spend, before taxes?

b) Did Justin spend enough
    money to avoid paying tax?
    If your answer is yes, how much
    more than $25 did Justin spend?
    If your answer is no, how much
    more would he need to spend
    and not pay the tax?

7. Write each fraction as a percent, then as a decimal.

a) $\frac{3}{4}$ b) $\frac{7}{25}$ c) $\frac{9}{10}$ d) $\frac{8}{200}$

8. This Chinese Yin Yang symbol is made from 5 circles. Suppose the radius of each medium-sized circle is 5 cm. What is the diameter of the largest circle? What assumptions did you make? Explain how you solved the problem.

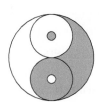

9. A car tire has radius about 29 cm.

a) What is the diameter of the tire?

b) Calculate the circumference of the tire.

c) How far has the car tire moved after one complete rotation? Give your answer to the nearest whole number.

d) About how many rotations will the tire make when the car travels 10 m?

10. Find the area of each shape.

a)

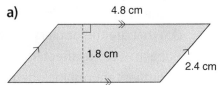

4.8 cm

1.8 cm

2.4 cm

b)

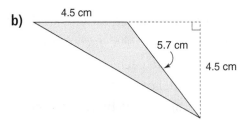

4.5 cm

5.7 cm

4.5 cm

11. Use a model to show each sum. Sketch the model. Write an addition equation for each picture.

a) $\frac{3}{5} + \frac{2}{10}$ b) $\frac{1}{3} + \frac{1}{12}$

c) $\frac{1}{4} + \frac{7}{8}$ d) $\frac{1}{4} + \frac{5}{6}$

12. A baker's cookie recipe calls for $6\frac{1}{8}$ cups of white sugar and $4\frac{1}{3}$ cups of brown sugar.

a) Estimate how much more white sugar is called for.

b) Calculate how much more white sugar is called for.

c) Draw a diagram to model your calculations in part b.

13. In a coin toss game, heads score $+1$ and tails score -1.

a) Write an equation you can use to solve each problem.

b) Solve the equation using tiles.

c) Verify each solution. Show your work.

  i) Meliq tossed a tail. He then had -2 points. How many points did Meliq have to begin with?

  ii) Vera tossed a head. She then had -3 points. How many points did Vera have to begin with?

14. Write an equation you could use to solve each problem. Solve each equation by systematic trial or by inspection.

a) Camille bought 9 teen magazines for $63. She paid the same amount for each magazine. How much did each magazine cost?

b) Nicolas collects fishing lures. He lost 27 of his lures on a fishing trip. Nicolas has 61 lures left. How many lures did he have to begin with?

15. Mary is a real estate agent in Lethbridge. One month she sold 7 houses at these prices: $171 000, $165 000, $178 000, $161 000, $174 000, $168 000, $240 000

a) Find the median price.

b) Do you think the mean price is greater than or less than the median price? Explain. Calculate to check.

c) What is the range of these prices?

16. Use these data: 28, 30, 30, 31, 32, 33, 34, 35, 37, 38, 39, 41

a) Find the mean, median, and mode.

b) What happens to the mean, median, and mode in each case?

i) Each number is increased by 10.

ii) Each number is doubled.

Explain the results.

17. The masses, in tonnes, of household garbage collected in a municipality each weekday in April are: 285, 395, 270, 305, 320, 300, 290, 310, 315, 295, 310, 295, 305, 325, 315, 310, 305, 300, 325, 305, 305, 300

a) Calculate the mean, median, and mode for the data.

b) What are the outliers? Explain your choice. Calculate the mean without the outliers. What do you notice? Explain.

c) When might you want to include the outliers? Explain.

18. This table shows the hourly wages of the employees at *Tea Break for You*.

Hourly Wage	Number of Employees
$7.50	4
$7.75	6
$8.00	3
$8.50	3
$8.75	2
$10.00	1
$12.50	1

a) Find the mean, median, and mode for these hourly wages.

b) Which measure best represents the wages? Explain.

c) What are the outliers? How is each average affected when the outliers are not included? Explain.

d) Who might earn the wages that are outliers? Explain.

19. Is this conclusion true or false? Explain.
The mean test score was 68%. Therefore, one-half the class scored above 68%.

20. Write the probability of each event as many different ways as you can.
a) Roll a 4 on a number cube labelled 1 to 6.
b) December immediately follows November.
c) Pick a red cube from a bag that contains 3 blue cubes, 4 green cubes, and 5 yellow cubes.

21. a) List the possible outcomes for rolling an octahedron labelled 1 to 8 and rolling a die labelled 1 to 6.
b) Why are the events in part a independent?
c) For how many outcomes are both numbers rolled less than 3?

22. Draw line segment MN.
Mark a point P not on MN.
Draw a line segment perpendicular to MN that passes through point P.

23. a) Draw line segment FG of length 7 cm. Use a ruler and compass to construct the perpendicular bisector of FG. Explain how you can check that the line you drew is the perpendicular bisector.

b) Draw ∠PQR = 140°. Use any method to bisect the angle. Use another method to check that the bisector you have drawn is correct.

24. Suppose you are given the coordinates of a point. You do not plot the point. How can you tell which quadrant the point will be in?

25. a) Plot these points: A(5, 10), B(−5, 10), C(−5, 0), D(−15, 0), E(−15, 10), F(−25, 10), G(−25, −20), H(−15, −20), J(−15, −10), K(−5, −10), L(−5, −20), M(5, −20)
b) Join the points in order. Then join M to A.
c) Explain how you chose the scale.
d) Describe the shape you have drawn.

26. A triangle has vertices C(−1, 5), D(3, 5), and E(3, −1).
a) Plot, then join, the points to draw △CDE.
b) Translate △CDE 2 units left and 4 units up. Write the coordinates of each vertex of the image △C′D′E′.
c) Reflect △C′D′E′ in the x-axis. Write the coordinates of each vertex of the image △C″D″E″.
d) Rotate △C″D″E″ 90° counterclockwise about the origin. Write the coordinates of each vertex of the image △C‴D‴E‴.

Unit 1 Patterns and Relations, page 4

1.1 Patterns in Division, page 8

1. Divisible by 2: parts a, c, and f
Divisible by 5: parts b, d, and f

2. Answers may vary. For example: a number with 0 in ones place is divisible by 2 and by 5. So, it is divisible by 10.

3. Divisible by 4: parts a, b, d, e, and f
Divisible by 8: parts b and f
Divisible by 10: parts c and d

4. Maxine is right. Tony is wrong. A number is divisible by 8 if, when divided by 4, the quotient is even (divisible by 2).

5. Answers may vary. For example: Multiples of 1000 are divisible by 8: 3000, 5000, 8000

6.a) Divisible by 2: 28, 54, 224, 322, 382, 460, 1046, 1088, 1784, 3662
Divisible by 4: 28, 224, 460, 1088, 1784
Divisible by 8: 224, 1088, 1784

c) Answers may vary. For example: 3472, 7000, 9632, all divisible by 8

7. Answers may vary. For example:
a) 0, 4, 8
b) 0, 2, 4, 6, 8
c) 0, 1, 2, 3, 4, 5, 6, 7, 8, 9

8. 1852, 1788, 1992, and 2004 are divisible by 4. Yes, 1964 is divisible by 4, so it is a leap year.

1.2 More Patterns in Division, page 11

1. Divisible by 3: parts a, b, c, d, e, and f
Divisible by 9: parts a, b, e, and f

2. Answers may vary. For example: 3102, 5100, 2010

3. a, b, c, e, f

4.a) 1, 2, 3, 5, 6, 10, 15, 25, 30, 50, 75, 150
b) 1, 5, 19, 95
c) 1, 3, 9, 13, 39, 117
d) 1, 2, 4, 5, 8, 10, 16, 20, 40, 80

5.

	Divisible by 9	Not divisible by 9
Divisible by 4	144, 252, 468	68, 120, 128, 424
Not divisible by 4	153	235, 361

6. 240

7.a) Answers may vary. For example: 135
b) 1, 3, 5, 9, 15, 27, 45, 135

c) 990; 135

8.a) 2, 5, 8
b) 0, 3, 6, 9
c) 1, 4, 7

9.a) 2 cereal bars
b) 4 cereal bars
c) 24 cereal bars cannot be divided among 0 groups.
d) A whole number cannot be divided among 0 groups.

Unit 1 Reading and Writing in Math: Writing to Explain Your Thinking, page 15

1. 25
2. 22 times
3.a) 41 tiles
b) The 9th term has 37 tiles.

1.3 Algebraic Expressions, page 18

1.a) 3, x, 2   **b)** 5, n, 0
c) 1, w, 3   **d)** 2, p, 4

2. $7p + 9$

3.a) $n + 6$ **b)** $8n$
c) $n - 6$ **d)** $\dfrac{n}{4}$

4.a) **i)** $20.00   **ii)** $32.00
b) 4t

5.a) $2n + 3$ **b)** $2(n - 5)$ **c)** $\dfrac{n}{7} + 6$
d) $28 - n$ **e)** $n - 28$

6.a) **i)** $n + 4$
     ii) $4 + n$
     iii) $n - 4$
     iv) $4 - n$

b) In parts i) and ii), the numerical coefficient, the variable, and the constant term are the same. So, the algebraic expressions are the same.
In parts iii) and iv), the numerical coefficients and the constant terms are different. So, the algebraic expressions are different.

7.a) 9   **b)** 12   **c)** 7   **d)** 2   **e)** 13   **f)** 12

8.a) 19   **b)** 3   **c)** 35   **d)** 18   **e)** 21   **f)** 4

9.a) $7 \times 8 + 9 \times 12$
b) $7x + 45$
c) 10 h

10.a) $n = 6$ **b)** $n = 4$ **c)** $n = 2$
d) $n = 3$ **e)** $n = 6$ **f)** $n = 40$

1.4 Relationships in Patterns, page 23

1.a) **i)** The term is twice the term number.
   ii) $2n$
  b) **i)** The term is 2 more than the term number.
   ii) $n + 2$
  c) **i)** The term is the term number multiplied by 8.
   ii) $8n$
  d) **i)** The term is 5 more than the term number.
   ii) $n + 5$

2.a) $3n$ **b)** $n + 2$
  c) $\frac{n}{2}$ **d)** $4n + 10$

3.a) $10n$
  b) $300.00

4.a) $4n$
  b) 48 cm
  c) Answers may vary. For example:
     i) perimeter of an equilateral triangle with side length s
     ii) perimeter of a regular octagon with side length t

5. Answers may vary. For example:
  a) Karin's brother is 5 years older than she is.
  b) Canoe rental is $15 for the first hour plus $2 per each additional hour.
  c) There are 3 candies per person and one left over.

6.a) $65.00; $110.00
  b) $9p + 20$
  c) $18p + 20$
  d) $9p + 40$
  e) Answers may vary. For example:
     The variable p represents any number.
     So, I can replace p to find the value of the algebraic expression for any particular value of the variable.

7.a) $e + 8$
  b) $13.00
  c) $e + 5$
  d) $10.00
  e) $3.00

8.a) $4n$
  b) $n + 6$
  c) $n - 1$

9.a) **i)** The term is double the term number plus one.
     ii) $2n + 1$

b) **i)** The term is two less than three times the term number.
   ii) $3n - 2$
  c) **i)** The term is three less than four times the term number.
   ii) $4n - 3$

1.5 Patterns and Relationships in Tables, page 27

1.a)

Input x	Output $2x$
1	2
2	4
3	6
4	8
5	10

The output is double the input.

b)

Input m	Output $10 - m$
1	9
2	8
3	7
4	6
5	5

The output is ten minus the input.

c)

Input p	Output $3x + 5$
1	8
2	11
3	14
4	17
5	20

The output is 5 more than 3 times the input.

2.a) $7n$
  b) $3n + 1$
  c) $2n - 1$

3. a)

Input n	Output $3n + 4$
1	7
2	10
3	13
4	16

b)

Input n	Output $4n + 3$
1	7
2	11
3	15
4	19

4.a) $3x + 2$
  b) $6x - 5$
  c) $5x + 3$

5.a) The pattern rule for the input is: Start at 5. Add 10 each time. The pattern rule for the output is: Start at 1. Add 2 each time. When the Input number increases by 10, the Output number increases by 2.

b)

Input x	Output
65	13
75	15
85	17

c) $\frac{x}{5}$ is related to x

Unit 1 Mid-Unit Review, page 29

1. Divisible by 4: parts a, c, d, and e
Divisible by 8: parts c and d

2. Divisible by 3: 54, 123, 3756
Divisible by 5: 85
Divisible by 3 and 5: 735, 1740, 6195

3.a) 1, 5, 17, 85
   b) 1, 2, 4, 8, 17, 34, 68, 136
   c) 1, 2, 3, 5, 6, 9, 10, 15, 18, 27, 30, 45, 54, 90, 135, 270

4.a) $n + 7$
   b) $11n$
   c) $\frac{n}{6}$
   d) $4n - 3$
   e) $2 + 5n$

5.a) i) 15
      ii) 16
   b) i) 48
      ii) 1
   c) i) 6
      ii) 8
   d) i) 22
      ii) 18

6.a) i) The term is the term number multiplied by 6.
      ii) $6n$
   b) i) The term is 4 more than the term number.
      ii) $n + 4$

7.a) $12 + 2t$
   b) $32.00; $52.00
   c) $12 + 4t$

8.a) $4x + 3$
   b) $8x - 3$

1.6 Graphing Relations, page 33

1.a) Output: 4, 8, 12, 16, 20
   b) Output: 4, 5, 6, 7, 8
   c) Output: 10, 14, 18, 22, 26

3.a) Output: 8, 20, 32, 44, 56
   b) One square represents 4 units.
   c) The graph shows a linear relation: When the Input number increases by 2, the Output number increases by 12.

4.a) 10
   b) 5
   c) 24

d) Answers may vary. For example: At a bowling alley, shoe rental is $8 and lane rental is $2/h.

5.a) $3n + 5$

   c)

Number of Go-Cart Rides	Total Cost ($)
0	5
1	8
2	11
3	14
4	17
5	20

   d) i) $23.00   **ii)** 8 rides

6.a) ii   **b)** iii   **c)** i

7.a) $75 - 5s$ **b)**

Week	Amount Owing
2	65
4	55
6	45
8	35
10	25

   c) The graph goes down to the right. When the number of weeks increases by 2, the amount owing decreases by $10.00.
   d) i) $10.00
      ii) After 15 weeks

8.a) Answers will vary. For example: Maya is paid a flat rate of $6 plus $5 for each item she sells.

   b)

Input n	Output $5n + 6$
0	6
1	11
2	16
3	21
4	26
5	31
6	36

   c) The graph goes up to the right. When the Input number increases by 1, the Output number increases by 5.
   d) Questions may vary. For example: What is the output when the input is 8? (*46*) What is the input when the output is 41? (*7*)

1.7 Reading and Writing Equations, page 36

1.a) $n + 8 = 12$
   b) $n - 8 = 12$

2.a) Twelve more than a number is 19.
   b) Three times a number is 18.
   c) Twelve minus a number is 5.
   d) A number divided by 2 is 6.

3.a) $6p = 258$

  b) $\frac{s}{2} = 21$

  c) $6h = 36$

4. $4s = 156$

5. $p = 6 \times 9$

6.a) C

  b) D

  c) A

  d) B

7. $\frac{n}{4} + 10 = 14$

8.a) i) $5s = 295$

  ii) $7h = 28$

  iii) $2x + 20 = 44$

  iv) $n + 7 = 20$

  b) Answers may vary. For example:
  The equation in part iii is the most difficult
  because it involves more operations.

  c) Answers may vary. For example: One-third
  the number of books on my shelf is 6.

1.8 Solving Equations Using Algebra Tiles, page 41

1.a) $x = 7$

  b) $x = 8$

  c) $x = 4$

  d) $x = 8$

  e) $x = 6$

  f) $x = 3$

2.a) $x + 7 = 12$

  b) $x = 5$

3. Answers may vary. For example:

  a) 6 and 13, 1, x

  b) 4 and 12, 1, x

  c) 11 and 7, 1, x

  d) 16, 2, x

  e) 18, 3, x

  f) 12, 4, x

4.a) $3x = 12$

  b) $x = 4$

5.a) $4x = 20$

  b) $x = 5$

6.a) $13 + x = 20$

  b) $x = 7$

7.a) $3x + 4 = 16$

  b) $x = 4$

8.a) $4x + 2 = 18$

  b) $x = 4$

9.a) $3x + 5 = 20$

  b) $x = 5$

10. Answers may vary. For example:

  a) $3x + 2 = 14$

  b) Two more than three times a number is 14.

  c) $x = 4$

  d) Tina had $14. She bought boxes of cookies at
  $3 per box. How many boxes did she buy if
  she was left with $2?

Unit 1 Unit Review, page 44

1. 1, 2, 3, 5, 6, 9, 10, 15, 18, 30, 45, 90

2. Parts a, b, c, d, e, f, h

3. 252 and 432

4.a) Yes. There are numbers divisible by 6
  and by 9.

  b) Divisible by 6: 330, 858
  Divisible by 9: 639, 2295
  Divisible by 6 and 9: 5598, 12 006
  Divisible by neither 6 nor 9: 10 217, 187

5.a) i) $n - 5$

  ii) 3

  b) i) $n + 10$

  ii) 18

  c) i) $3n$

  ii) 24

  d) i) $3n + 6$

  ii) 30

6.a) $4n$

  b) $n + 3$

  c) $\frac{n}{4}$

7.a)

Input n	Output $n + 13$
1	14
2	15
3	16
4	17
5	18

b)

Input n	Output $5n + 1$
1	6
2	11
3	16
4	21
5	26

c)

Input n	Output $6n - 3$
1	3
2	9
3	15
4	21
5	27

8.a) $n + 11$

  b) $5n - 3$

9.a) iv

  b) i

  c) v

10. Answers may vary. For example:
  a) **i)** The cost is $4 plus $2/h.
  ii)

Input m	Output $4 + 2m$
1	6
2	8
3	10
4	12
5	14

  iv) The graph goes up to the right.
  When the Input number increases by 1, the Output number increases by 2.
  v) Questions may vary. For example:
  What is the input when the output is 18? (*7*)
  What is the output when the input is 6? (*16*)
  b) **i)** Anna owes her mother $15. She pays her $2/week.
  ii)

Input d	Output $15 - 2d$
0	15
1	13
2	11
3	9
4	7

  iv) The graph goes down to the right.
  When the Input number increases by 1, the Output number decreases by 2.
  v) What is the input when the output is 3? (*6*)
  What is the output when the input is 7? (*1*)
11.a) $2c + 6$
  b)

c	Amount Paid ($)
0	6
5	16
10	26
15	36

  c) The graph goes up to the right.
  When the number of children supervised increases by 5, the amount paid increases by $10.00.
  d) **i)** $56.00
  ii) 20 children
12. Answers will vary. For example:
  May is payed $24 per day, plus $2 for each dress she sells.
13.a) $3n = 15$

b) $3n - 4 = 20$
14. $8n = 48$
15.a) **i)** $3x = 36$
  ii) $x = 12$
  b) **i)** $x + 7 = 18$
  ii) $x = 11$
  c) **i)** $3x = 24$
  ii) $x = 8$
  d) **i)** $x + 8 = 21$
  ii) $x = 13$
16.a) $4x + 5 = 21$
  b) $x = 4$

Unit 1 Practice Test, page 47
  1.a) 0, 2, 4, 6, 8
  b) 2, 5, 8
  c) 2, 6
  d) 0, 5
  e) 2, 8
  f) 6
  g) 8
  h) 0
2. For $n = 1$, $2 + 3n$ equals $2n + 3$.
  For $n = 5$, $2n + 3$ equals $3n - 2$.
3.a) $25 + 2v$
  b) $45.00; $75.00
  c) $25 + 3v$; Jamal would pay $55.00; that is, $10.00 more.
4.a) **i)** $x + 5 = 22$
  ii) $2x = 14$
  iii) $3x + 4 = 19$
  b) **i)** $x = 17$
  ii) $x = 7$
  iii) $x = 5$

Unit 2 Integers, page 50

2.1 Representing Integers, page 54
1.a) $+1$ **b)** $+3$ **c)** 0 **d)** -1 **e)** -3 **f)** -2
2. Answers may vary. For example:
  a) 6 red tiles, or 7 red tiles and 1 yellow tile
  b) 7 yellow tiles, or 8 yellow tiles and 1 red tile
  c) 6 yellow tiles, or 8 yellow tiles and 2 red tiles
  d) 2 red tiles, or 6 red tiles and 4 yellow tiles
  e) 9 yellow tiles, or 10 yellow tiles and 1 red tile
  f) 4 red tiles, or 5 red tiles and 1 yellow tile
  g) 1 yellow tile and 1 red tile, or 3 yellow tiles and 3 red tiles
  h) 10 yellow tiles, or 13 yellow tiles and 3 red tiles

3.a)

Number of Yellow Tiles	Number of Red Tiles	Integer Modelled
0	6	–6
1	5	–4
2	4	–2
3	3	0
4	2	+2
5	1	+4
6	0	+6

4.a) I chose +3. I need 3 yellow tiles to model it.

b) I add a zero pair each time. I can model +3 in many ways.

c)

Number of Yellow Tiles	Number of Red Tiles	Integer Modelled
3	0	+3
4	1	+3
5	2	+3
6	3	+3

There are always 3 more yellow tiles than red tiles. As the number of yellow tiles increases, the number of red tiles increases by the same amount.

d) For a negative integer, such as –23, there will always be 23 more red tiles than yellow tiles. For a positive integer, such as +41, there will be 41 more yellow tiles than red tiles.

5.a) 8   **b)** 98

6.a) +9   **b)** –5   **c)** +11   **d)** –9   **e)** –7

7.a) +100; –20

  b) +6; –4

  c) +12; –8

2.2 Adding Integers with Tiles, page 58

1.a) $(+4) + (-2) = +2$

  b) $(+2) + (-3) = -1$

  c) $(-4) + (-2) = -6$

  d) $(+6) + (-3) = +3$

  e) $(+1) + (-4) = -3$

  f) $(+3) + (+2) = +5$

2.a) +1   **b)** –1   **c)** 0

3.a) 0   **b)** 0   **c)** 0

The number of red tiles equals the number of yellow tiles each time.

4.a) +5   **b)** +1   **c)** –5

5.a) $(+4) + (+3) = +7$

  b) $(-7) + (+5) = -2$

  c) $(-4) + (-5) = -9$

  d) $(+8) + (-1) = +7$

  e) $(-10) + (-6) = -16$

  f) $(+4) + (-13) = -9$

6.a) $(-3) + (+4) = +1$

  b) $(+5) + (-3) = +2$

  c) $(+15) + (-7) = +8$

  d) $(-3) + (+8) = +5$

  e) $(+12) + (-5) = +7$

8.a) (+3)   **b)** (–1)   **c)** (–2)

  d) (+2)   **e)** (–1)   **f)** (+6)

9.a) –4

  b) No, the sum remains the same.

  c) Each integer has been replaced by its opposite. The sum is also replaced by its opposite.

10.a) +6   **b)** +4   **c)** –5   **d)** +2

11. a)

+3	–4	+1
–2	0	+2
–1	+4	–3

b)

–1	–6	+1
0	–2	–4
–5	+2	–3

12.a) –8, –12, –16, –20 …

  Add –4 each time to get the next term.

  b) 0, +3, +6, +9 …

  Add +3 each time to get the next term.

2.3 Adding Integers on a Number Line, page 62

1.a) +4   **b)** +2   **c)** –2   **d)** –4

  e) –7   **f)** +1   **g)** –1   **h)** +7

2.a) +6   **b)** +2   **c)** –6   **d)** –6

  e) –13   **f)** –5   **g)** –3   **h)** +12

3. a), b) The answers are the same.

  c) The order in which you add integers does not matter.

4.a) –2   **b)** –3   **c)** +4

5.a) +5; The temperature rose 5°C.

  b) +4; Adrian gained $4.

  c) +1; The stock was up $1.

6.a)   **i)** –2

    ii) +5

    iii) –6

    iv) +8

  b)   **i)** $(+2) + (-2) = 0$

    ii) $(-5) + (+5) = 0$

    iii) $(+6) + (-6) = 0$

    iv) $(-8) + (+8) = 0$

  c) The sum of two opposite integers is 0.

7. a), b) i) $(-5) + (-10) = -15$

    You take 15 steps backward.

    ii) $(-5) + (+8) = +3$;

    You deposit $3.

    iii) $(-8) + (+6) = -2$;

    The diver descends 2 m.

iv) $(+4) + (-7) = -3$;
The snowmobile driver rides
3 km west.
v) $(+6) + (-10) = -4$;
The person loses 4 kg.

8.a) i) $(-4) + (+7) = +3$
ii) $(+8) + (-3) = +5$

b) Answers may vary. For example:
i) The temperature dropped 4°C overnight and rose 7°C during the day.
ii) Sarah has $8 and spends $3.

9.a) Always true
b) Never true
c) Always true
d) Sometimes true

10.a) +1 **b)** −5 **c)** −6 **d)** 0

11. +6°C

Unit 2 Mid-Unit Review, page 65

1. Answers may vary. For example:
a) 5 red tiles, or 6 red tiles and 1 yellow tile
b) 1 red tile and 1 yellow tile, or 4 red tiles and 4 yellow tiles
c) 8 yellow tiles, or 9 yellow tiles and 1 red tile
d) 3 red tiles and 2 yellow tiles, or 1 red tile
e) 3 yellow tiles, or 4 yellow tiles and 1 red tile
f) 7 red tiles, or 9 red tiles and 2 yellow tiles

2. 11

3.a) +5 **b)** −2 **c)** 0

4.a) +3 **b)** −5 **c)** −4 **d)** +9 **e)** −12 **f)** +12

5.a) +5 **b)** −6 **c)** −2 **d)** +1 **e)** 0 **f)** +7

6.a) −1
b) Answers may vary. For example:
+2 and −3; +3 and −4; +5 and −6; +6 and −7

7.a) $(+50) + (-20) = +30$;
Puja had $30.
b) $(+5) + (-10) = -5$;
The temperature was −5°C.
c) $(+124\,000) + (-4000) = +120\,000$;
The population was 120 000.
d) $(+12\,000) + (-1200) = +10\,800$;
The plane was cruising at 10 800 m.

8.a) i) $(-2) + (+6) = +4$
ii) $(+4) + (-6) = -2$
b) Answers may vary. For example:
i) The temperature was −2°C and it rose 6°C.
ii) Karin walked 4 steps forward and 6 steps backward.

9.a) $(+1) + (+2) + (+3) + (+4) = +10$
b) $(-1) + (0) + (+1) = 0$ or
$(-2) + (-1) + (0) + (+1) + (+2) = 0$

c) $(-1) + (0) + (+1) + (+2) = +2$
d) $(+3) + (+4) = +7$
e) $(-3) + (-2) + (-1) + (0) + (+1) + (+2) + (+3) + (+4) = +4$
f) $(-7) + (-6) + (-5) + (-4) + (-3) + (-2) + (-1) + (0) + (+1) + (+2) + (+3) + (+4) + (+5) + (+6) + (+7) + (+8) = +8$

2.4 Subtracting Integers with Tiles, page 69

1.a) +3 **b)** 0 **c)** −3
d) +2 **e)** −7 **f)** 0
2.a) +3 **b)** −5 **c)** +7
d) −1 **e)** +2 **f)** −9
3.a) −3 **b)** +5 **c)** −7
d) +1 **e)** −2 **f)** +9
4.a) +11 **b)** −10 **c)** −14
d) +14 **e)** −9 **f)** −12
5.a) −1 **b)** −8 **c)** −7
d) +7 **e)** +10 **f)** +11

7.a) i) +2 and −2
ii) −1 and +1
iii) +7 and −7
b) When the order in which we subtract two integers is reversed, the answer is the opposite integer.

8. −7

9. I can write as many questions as I want.
For example:
a) $(-4) - (-6) = +2$
$(+7) - (+5) = +2$
$(+1) - (-1) = +2$
b) $(-5) - (-2) = -3$
$(-4) - (+7) = -3$
$(-1) - (+2) = -3$
c) $(-3) - (-8) = +5$
$(+7) - (+2) = +5$
$(+2) - (-3) = +5$
d) $(-8) - (-2) = -6$
$(+3) - (+9) = -6$
$(-3) - (+3) = -6$

10.a) Part i; +4 is greater than −4.
b) Part i; +1 is greater than −1.

11.a) +2 and −3
b) Answers will vary. For example:
Find two integers with a sum of +3 and a difference of +9. Answer: +6 and −3

12.a) (+1) **b)** (+4) **c)** (+5)

13.a) +2 **b)** 0 **c)** 0 **d)** +1 **e)** −3 **f)** 0

14.a) The sum of the numbers in each row, column, and diagonal is –9, so the square is still magic.

–4	+1	–6
–5	–3	–1
0	–7	–2

b) The sum of the numbers in each row, column, and diagonal is +6, so the square is still magic.

+1	+6	–1
0	+2	+4
+5	–2	+3

2.5 Subtracting Integers on a Number Line, page 73

1.a) +1 **b)** +7 **c)** –3 **d)** –7 **e)** +4 **f)** +4

2.a) –1, –7, +3, +7, –4, –4

b) The answers in part a are the opposites of those in question 1. When the order of the integers is reversed, the difference changes to its opposite.

3.a) +5   **b)** +10   **c)** –14
  d) –15   **e)** –8   **f)** 0

4.a) $(+6) + (-4) = +2$
  b) $(-5) + (-4) = -9$
  c) $(-2) + (+3) = +1$
  d) $(+4) + (+2) = +6$
  e) $(+1) + (-1) = 0$
  f) $(+1) + (+1) = +2$

5.a) +12°C or –12°C
  b) +7°C or –7°C
  c) +13°C or –13°C

6.a) +8 or –8
  b) +5 or –5
  c) +9 or –9

7.a) **i)** $(+13) – (-4) = +17; +17°C$
    ii) $(-10) – (-22) = +12; +12°C$
    iii) $(+12) – (-3) = +15; +15°C$
    iv) $(+13) – (+7) = +6; +6°C$
  b) Calgary

8.a) –17
  b) +17; the answers in parts a and b are opposite integers.
  c) Each integer was replaced with its opposite. The differences are opposite integers: +17 and –17

9. Answers may vary. For example:
  $(-6) – (-10) = +4$
  $(+6) – (+2) = +4$
  $(-1) – (-5) = +4$

10.a) $(+6) – (+5) = +1$
  $(+5) – (+5) = 0$
  $(+4) – (+5) = -1$
  $(+3) – (+5) = -2$
  $(+2) – (+5) = -3$
  b) $(+7) – (+4) = +3$
  $(+7) – (+3) = +4$
  $(+7) – (+2) = +5$
  $(+7) – (+1) = +6$
  $(+7) – (0) = +7$
  $(+7) – (-1) = +8$
  $(+7) – (-2) = +9$
  $(+7) – (-3) = +10$
  c) $(+8) – (+7) = +1$
  $(+7) – (+7) = 0$
  $(+6) – (+7) = -1$
  $(+5) – (+7) = -2$
  $(+4) – (+7) = -3$
  $(+3) – (+7) = -4$
  $(+2) – (+7) = -5$
  $(+1) – (+7) = -6$
  $0 – (+7) = -7$
  $(-1) – (+7) = -8$
  $(-2) – (+7) = -9$
  $(-3) – (+7) = -10$

11.a) –6, –10, –14, –18;
    Start at +6. Subtract +4 each time.
  b) +3, +5, +7, +9;
    Start at –3. Subtract –2 each time.
  c) +26, +33, +40, +47;
    Start at +5. Subtract –7 each time.
  d) –2, –3, –4, –5;
    Start at +1. Subtract +1 each time.

12.a) +1 **b)** +1 **c)** –4 **d)** +2 **e)** +12 **f)** –11

Unit 2 Unit Review, page 79

1.a) 5   **b)** 17   **c)** 37   **d)** 0
2.a) +8 **b)** –5 **c)** +12 **d)** –7 **e)** –9
3.a) –3   **b)** +1   **c)** –1   **d)** 0
4.a) $(-6) + (+4) = -2$
  b) $(-25) + (+13) = -12$
  c) $(+15) + (-23) = -8$
  d) $(-250) + (+80) = -170$
5. Answers may vary. For example:
  a) $(-5) + (0) = -5;$
    $(-3) + (-2) = -5;$
    $(-1) + (-4) = -5;$
    $(+1) + (-6) = -5$

b) $(+4) + (0) = +4$;
$(+2) + (-2) = +4$;
$(-2) + (+6) = +4$;
$(-4) + (+8) = +4$
6. $(-10) + (+17) = +7$;
The new temperature is $+7°C$.
7.a) i) $(-4) + (+5) = +1$
ii) $(+2) + (-4) = -2$
b) Answers may vary. For example:
i) Sasha takes 4 steps backward and 5 steps forward.
ii) The temperature is $+2°C$ and then drops $4°C$.
8.a) $+2$ **b)** -1 **c)** -5 **d)** $+2$
9.a) $+2$ **b)** $+2$ **c)** -10 **d)** -2
10. The difference of two positive integers is positive if the first integer is greater than the second integer. The difference of two positive integers is negative if the first integer is less than the second integer.
11.a) $+9°C$ **b)** $0°C$ **c)** $-6°C$ **d)** $-7°C$
12.a) $+3$ **b)** $+6$ **c)** $+4$ **d)** -5
e) -4 **f)** -5 **g)** -2 **h)** $+5$
13.a) $+5$ **b)** -10 **c)** $+1$ **d)** 0 **e)** $+6$ **f)** -1
14.a) $+12°C$ or $-12°C$
b) -150 m or $+150$ m
15.a) -9 m or $+9$ m
b) $+14$ m or -14 m
16.a) $+12$ kg or -12 kg
b) -1 kg or $+1$ kg
17.a) $+1$ **b)** -2 **c)** $+3$ h or -3 h
18. Answers may vary. For example:
a) $(+10) - (+4) = +6$
$(+8) - (+2) = +6$
$(+6) - (0) = +6$
$(+4) - (-2) = +6$
$(+2) - (-4) = +6$
b) $(-5) - (-2) = -3$
$(-1) - (+2) = -3$
$(+3) - (+6) = -3$
$(0) - (+3) = -3$
$(-3) - (0) = -3$

Unit 2 Practice Test, page 81
1.a) -3 **b)** -10 **c)** -10
d) $+6$ **e)** -4 **f)** $+23$
2.a) $+8$ **b)** -15 **c)** -11
d) $+7$ **e)** $+2$ **f)** $+4$
3.a) The sum of two integers is zero when the integers are opposites.
b) The sum of two integers is negative when both integers are negative; or when one

integer is positive and the other is negative, and the negative integer has a longer arrow on the number line.
c) The sum of two integers is positive when both integers are positive; or when one integer is positive and the other is negative, and the positive integer has a longer arrow on the number line.
4.a) 6 different scores
b) $(+10) + (+10) = +20$
$(+10) + (+5) = +15$
$(+10) + (-2) = +8$
$(+5) + (+5) = +10$
$(+5) + (-2) = +3$
$(-2) + (-2) = -4$
5. $+373°C$ or $-373°C$
6. There are 4 possible answers: $+7$, $+13$, -1, and $+5$.
For 4 integers in a row, the addition and/or subtraction signs can be arranged as shown:
$+ + +$; $+ + -$; $+ - +$; $+ - -$; $- + +$; $- + -$; $- - +$; $- - -$

Unit 2 Unit Problem: What Time Is It?, page 82
1.a) 0:00 a.m.
b) 5:00 a.m.
c) 9:00 a.m.
d) 6:00 a.m.
2. 10:00 a.m. the next day
3. Atsuko needs to fly out at 3:00 p.m. Tokyo time.
Paula needs to fly out at 7:00 a.m. Sydney time.

Unit 3 Fractions, Decimals, and Percents, page 84

3.1 Fractions to Decimals, page 88
1.a) i) $0.\overline{6}$
ii) 0.75
iii) 0.8
iv) $0.8\overline{3}$
v) $0.\overline{857\,142}$
b) i) repeating
ii) terminating
iii) terminating
iv) repeating
v) repeating
2.a) $\dfrac{9}{10}$

b) $\frac{26}{100} = \frac{13}{50}$

c) $\frac{45}{100} = \frac{9}{20}$

d) $\frac{1}{100}$

e) $\frac{125}{1000} = \frac{1}{8}$

3. a) i) $0.\overline{037}$

    ii) $0.\overline{074}$

    iii) $0.\overline{1}$

b) As the numerator of the fraction increases by 1, the corresponding decimal increases by $0.\overline{037}$ each time.

c) i) $0.\overline{148}$

    ii) $0.\overline{185}$

    iii) $0.\overline{296}$

4.a) $\frac{4}{10}$, 0.4

b) $\frac{25}{100}$, 0.25

c) $\frac{52}{100}$, 0.52

d) $\frac{38}{100}$, 0.38

e) $\frac{74}{1000}$, 0.074

5.a) $\frac{2}{3}$ **b)** $\frac{5}{9}$ **c)** $\frac{41}{99}$ **d)** $\frac{16}{99}$

6. a) $0.\overline{571\,428}$ **b)** $0.\overline{4}$

  c) $0.5\overline{4}$ **d)** $0.\overline{538\,461}$

7. $0.294\,117\,647$; Use long division.

8. 0.2

  a) 0.8 **b)** 1.4 **c)** 1.8 **d)** 2.2

9.a) i) $0.\overline{001}$

    ii) $0.\overline{002}$

    iii) $0.\overline{054}$

    iv) $0.\overline{113}$

b) The numerator of the fraction becomes the repeating digits in the decimal. If the numerator is a two-digit number, the first repeating digit is 0.

c) i) $\frac{4}{999}$

    ii) $\frac{89}{999}$

    iii) $\frac{201}{999}$

iv) $\frac{326}{999}$

10.a) iii   **b)** i   **c)** iv   **d)** ii

11.a) $1.0, 2.0, 1.5, 1.\overline{6}, 1.6, 1.625$; The decimals are greater than or equal to 1 and less than or equal to 2.

  b) $1.\overline{615\,384}$, $1.\overline{619\,047}$, $1.617\,647...$, $1.\overline{618}$

12.a) $1.\overline{142\,857}$; Six digits repeat.

  b) $0.\overline{285\,714}$, $0.\overline{428\,571}$, $0.\overline{571\,428}$, $0.\overline{714\,285}$, $0.\overline{857\,142}$; The tenth digit increases from least to greatest; the other digits follow in a clockwise direction around the circle.

13.a) i) 0.875; terminating

    ii) $0.2\overline{7}$; repeating

    iii) 0.3; terminating

    iv) $0.\overline{296}$; repeating

    v) 0.16; terminating

  b) i) $2 \times 2 \times 2$

    ii) $2 \times 3 \times 3$

    iii) 2×5

    iv) $3 \times 3 \times 3$

    v) 5×5

  c) When the prime factors of the denominator are 2 and 5 only, the corresponding decimal is terminating. When the denominator has any other prime factors, the fraction can be written as a repeating decimal.

  d) i) No

    ii) Yes

    iii) No

    iv) Yes

3.2 Comparing and Ordering Fractions and Decimals, page 94

1. Answers may vary.

For example: $\frac{1}{7}$, $\frac{4}{7}$, $\frac{8}{7}$, $\frac{18}{7}$, $\frac{24}{7}$

2. From greatest to least: $\frac{11}{3}$, $2\frac{5}{6}$, $2\frac{1}{2}$

3.a) 1, $\frac{7}{6}$, $1\frac{2}{9}$, $\frac{15}{12}$

  b) $\frac{7}{6}$, $1\frac{3}{4}$, 2, $\frac{7}{3}$

  c) $\frac{15}{10}$, $\frac{7}{4}$, 2, $\frac{11}{5}$

  d) $2\frac{1}{3}$, $\frac{10}{4}$, 3, $\frac{9}{2}$

4.a) $3\frac{1}{2}$, $\frac{13}{4}$, $3\frac{1}{8}$; $3.5, 3.25, 3.125$

b) $1\frac{1}{12}, \frac{5}{6}, \frac{9}{12}, \frac{2}{3}$; $1.08\overline{3}, 0.8\overline{3}, 0.75, 0.\overline{6}$

c) $\frac{3}{2}, 1\frac{2}{5}, \frac{4}{3}$; $1.5, 1.4, 1.\overline{3}$

5.a) $1, 1.25, 1.6, \frac{7}{4}, 1\frac{4}{5}$

b) $1.875, 2, \frac{5}{2}, 2\frac{5}{8}, 2\frac{3}{4}$

6.a) $\frac{17}{5}, 3\frac{1}{4}, 3.2, \frac{21}{7}, 2.8, 2$

7. Answers may vary. For example:

a) $\frac{27}{16}$ **b)** 2.25

8. Answers may vary. For example:

a) $\frac{11}{14}$ **b)** $1\frac{1}{2}$ **c)** 1.35 **d)** 0.55

9.a) $\frac{11}{4}$; $2\frac{1}{2} = \frac{10}{4}$ which is less than $\frac{11}{4}$.

b) $3\frac{2}{5}$; $\frac{2}{5}$ is close to $\frac{1}{2}$, so $3\frac{2}{5}$ is closer to $3\frac{1}{2}$.

10.a) $6\frac{2}{20}$ should be the second number in the set:

$\frac{29}{5}, 6\frac{2}{20}, 6\frac{2}{10}, 6.25$

b) $\frac{3}{2}$ should be the first number in the set:

$\frac{3}{2}, 1\frac{7}{16}, 1\frac{3}{8}, 1.2, \frac{3}{4}$

11.a) From least to greatest: $\frac{11}{6}, 1.875, \frac{9}{4}$

b) Corey sold the most pizzas; Amrita sold the fewest pizzas.

c) Use equivalent fractions.

d) $\frac{11}{6}, 1.875, 2\frac{1}{5}, \frac{9}{4}$

3.3 Adding and Subtracting Decimals, page 98

1.a) $2 - 0 = 2$

b) $71 + 6 = 77$

c) $125 + 37 = 162$

d) $9 - 1 = 8$

2. 0.067 km

3.a) $819.24

b) $248.26

4. a) 12.7 kg

b) No; 12.7 is greater than 10.5.

c) 2.2 kg

5. Use front-end estimation: 49; 51.485

6.a) Robb family: $428.79; Chan family: $336.18

b) $92.61

7. Answers may vary.
For example: 216.478 and 65.181

8. Answers may vary.
For example: 0.312 and 5.476

9.a) The student did not line up the digits of like value.

b) 4.437

10. Answers may vary.
For example: 1.256 and 2.044

11.a) Start at 2.09. Add 0.04 each time.

b) Start at 5.635. Subtract 0.25 each time.

3.4 Multiplying Decimals, page 102

1.a) $1.7 \times 1.5 = 2.55$

b) $2.3 \times 1.3 = 2.99$

2.a) 3.9

b) 0.92

c) 0.56

3. Answers may vary. For example: I chose part a from question 2. I used 2 flats: $2 \times 1 = 2$;
16 rods: $16 \times 0.1 = 1.6$;
30 small cubes: $30 \times 0.01 = 0.3$.
The area of the plot is: $2 + 1.6 + 0.3 = 3.9$

4.a) 15.54 **b)** 2.67 **c)** 0.54

5. 161.65; I estimated 150, so the answer is reasonable.

6.a) 83.6; 836; 8360; 83 600; Multiply by multiples of 10. The digits in the product move one place to the left each time. Or, the decimal point moves one place to the right.

b) 0.836; 0.0836; 0.008 36; 0.000 836; Multiply by multiples of 0.1. The digits in the product move one place to the right each time. Or, the decimal point moves one place to the left.

7. 9.18 m^2

8.a) 12.922 2

b) 174.315 96

c) 1.333 072

9.a) 936.66 km

b) 852.24 km

10.a) $2.43 **b)** $12.50 **c)** $0.62

11. Answers may vary.
For example: 1.2 and 0.3 or 0.2 and 1.8

12.a) 216

b) **i)** 21.6

ii) 2.16

iii) 2.16

iv) 0.0216

13.a) **i)** 11.34

ii) 0.0962

iii) 8.448

iv) 1.1106

b) The number of decimal places in the product is the sum of the number of decimal places in the question.

c) 9.1; Yes, the rule applies, but the product must be written as 9.10. The calculator does not show the product this way.

3.5 Dividing Decimals, page 106

1.a) 8   **b)** 4   **c)** 4.5   **d)** 5.5

2.a) 12.45; 1.245; 0.1245; 0.012 45; Divide by multiples of 10. The digits in the quotient move one place to the right each time. Or, the decimal point moves one place to the left.

b) 1245; 12 450; 124 500; 1 245 000; Divide by multiples of 0.1. The digits in the quotient move one place to the left each time. Or, the decimal point moves one place to the right.

3. All division statements are equivalent.

4.a) 11.9   **b)** 976.5   **c)** 39.15

5.a) 2.5   **b)** 3.2   **c)** 1.6   **d)** 2.4

6.a) 3.5   **b)** 1.5   **c)** 7.1   **d)** 24.1

7. 87

8. 27.9 m

9.a) About $3

b) $3.35

c) About 3 kg

10.a) About 12 pieces; Assumptions may vary.

b) No, he needs 14 pieces and he has material for 12.

c) If Alex cannot use the 0.28-m piece left after he cut twelve 0.8-m pieces, he needs 1.6 m of fabric. If he can use it, he only needs 1.32 m of fabric.

d) Yes; Alex would only need 0.7 m × 14 = 9.8 m of fabric.

11. Answers may vary.
For example: 0.312 and 2.6

12. $9.25; The result should be written to the nearest hundredth.

13. 237 is greater than 10 times 7 and less than 100 times 7, so the quotient should be between 10 and 100: 237 ÷ 7 = 33.857

a) 338.57 **b)** 33.857 **c)** 3.3857 **d)** 33.857

3.6 Order of Operations with Decimals, page 109

1.a) 6.5   **b)** 6.2   **c)** 14   **d)** 1498

2.a) 58   **b)** 211   **c)** 12

3.a) 4.4

b) 2.2

4.a) 345.68 **b)** 18.038

c) 163   **d)** 116.54

5.a) Aida

b) Ioana: $12 \times (4.8 \div 0.3 - 3.64 \times 3.5) = 39.12$
Norman: $(12 \times 4.8 \div 0.3 - 3.64) \times 3.5 = 659.26$

6. 41.21

7. Answers may vary. For example:
$0.1 + 0.2 + 0.3 + 0.4 = 1$,
$(0.6 \times 0.5 + 0.7) \times 0.2 \div 0.1 = 2$,
$(0.8 + 0.7) \times 0.6 \div 0.3 = 3$,
$0.6 \div 0.2 + 0.1 + 0.9 = 4$,
$0.9 \div 0.3 + 0.4 \div 0.2 = 5$

Unit 3 Mid-Unit Review, page 110

1.a)   **i)** $0.0\overline{3}$

    ii) $0.0\overline{6}$

    iii) $0.0\overline{9}$

b) Start at $0.0\overline{3}$. Add $0.0\overline{3}$ each time.

c)   **i)** $\dfrac{5}{33}$

    ii) $\dfrac{12}{33}$

2.a) 0.125; terminating

b) 0.6; terminating

c) $0.\overline{6}$; repeating

d) $0.\overline{538\,461}$; repeating

3.a) $\dfrac{1}{5}$ **b)** $\dfrac{8}{9}$ **c)** $\dfrac{1}{200}$ **d)** $\dfrac{23}{99}$

4. From least to greatest:

a) $\dfrac{11}{6}$, 2, $2\dfrac{1}{4}$, $\dfrac{8}{3}$ **b)** $1\dfrac{3}{4}$, $\dfrac{23}{8}$, 3.5

c) 1, $\dfrac{13}{10}$, $1\dfrac{3}{5}$, 1.75, $\dfrac{9}{5}$

5. Answers may vary. For example:

a) 1.5   **b)** 2.4   **c)** 1.5

6.a) 25.72   **b)** 137.521   **c)** 17.1

7.a) 3.585 kg   **b)** 9.25 kg

8.a) 7.44   **b)** 4.706   **c)** 58.95

9. 9.94 km^2

10. The division statements are equivalent.

11.a) 16.26   **b)** 50.5   **c)** 18.431

3.7 Relating Fractions, Decimals, and Percents, page 112

1.a) $\dfrac{3}{20}$, 15%, 0.15

b) $\dfrac{2}{5}$, 40%, 0.4

c) $\dfrac{4}{5}$, 80%, 0.8

2.a) $\dfrac{1}{50}$, 0.02

b) $\frac{9}{100}$, 0.09

c) $\frac{7}{25}$, 0.28

d) $\frac{19}{20}$, 0.95

3.a) 0.2, 20%
  b) 0.06, 6%
  c) 0.16, 16%
  d) 0.65, 65%
  e) 0.8, 80%
4. Janet; 82% is greater than 80%.
5. 15%
6.a) 25%   **b)** 50%   **c)** 6%   **d)** 10%

3.8 Solving Percent Problems, page 115
1.a) 3   **b)** 10   **c)** 6.48   **d)** 75.04
2.a) $45.00 **b)** $42.00 **c)** $36.00
3. a) $40.50 **b)** $22.00 **c)** $35.00
4. a) $3.63 **b)** $11.30 **c)** $3.27
5.a) **i)** $7.74
     ii) $136.74
  b) **i)** $1.50
     ii) $26.49
  c) **i)** $2.58
     ii) $45.55
6. About 192 bands
7.a) Answers may vary. For example: Some items will be 60% off, others will be reduced by less. Or, the sale prices will be at least 40% the original price.
  b) Scarves and hats
  c) Sweaters: About $20.00 ($14.99 off sale price), ski jackets: $60.00 ($52.49 off sale price), leather gloves: $28.00 ($10.49 off sale price)
8.a) $199.99 − $199.99 × 0.25 = $149.99
  b) $199.99 × 0.75 = $149.99
  c) Yes

Unit 3 Unit Review, page 121
1.a) 0.6; terminating
  b) $0.8\overline{3}$; repeating
  c) 0.375; terminating
  d) 0.15; terminating
2.a) $\frac{11}{20}$ **b)** $1\frac{1}{3}$ **c)** $\frac{4}{5}$ **d)** $\frac{7}{99}$
3.a) From least to greatest:
   $\frac{3}{6}$, $\frac{5}{8}$, $1\frac{1}{16}$, 1.1, $\frac{5}{4}$

4. For example:
  a) 2.25; From least to greatest:
   2.25, $2\frac{1}{3}$, $\frac{17}{6}$, $2\frac{11}{12}$
  b) $1\frac{3}{15}$; From least to greatest:
   $\frac{3}{5}$, $\frac{9}{10}$, $\frac{21}{20}$, 1.1, $1\frac{3}{15}$
5. Answers will vary.
  For example: 1.78 and 1.63
6. 0.72 s
7.a) $118.58
  b) $59.29
8. $1.56
9. i) a, b, c
  ii) d, e, f; part d: 4.1875; part e: 5.2; part f : 24.2
10. 6.25 m
11.a) 43.79
  b) 5.855
12.a) i) 10.68
    ii) 10.92
   iii) 9.48
   iv) 11.56
  b) When the position of the brackets changes, the order of operations changes.
13.a) $\frac{4}{5}$, 0.8
  b) $\frac{3}{25}$, 0.12
  c) $\frac{1}{50}$, 0.02
  d) $\frac{63}{100}$, 0.63
14.a) 0.56, 56%
  b) 0.95, 95%
  c) 0.14, 14%
  d) 0.2, 20%
15. 28 students
16.a) $33.15   **b)** $21.75   **c)** $31.50
17.a) $34.19   **b)** $31.79   **c)** $2.40
18. $6.55

Unit 3 Practice Test, page 123
1.a) $\frac{1}{250}$ **b)** $\frac{16}{25}$ **c)** $\frac{1}{3}$
  d) 0.255   **e)** 0.75
2.a) $90.00
  b) No. The equipment costs $107.80.
  c) $17.80
3. Yes
4.a) 34.74   **b)** 15.67

5. 26 cats
6.a) $58.50 **b)** $19.50 **c)** $3.51 **d)** $62.01

Cumulative Review Units 1–3, page 126

1. Divisible by 4: 320, 488, 2660
Divisible by 6: 762, 4926
Divisible by 4 and by 6: 264, 504
Not divisible by 4 or by 6: 1293
2.a) 5 strawberries   **b)** 8 strawberries
  c) I cannot divide 40 strawberries among
     0 people.
3.a) $\frac{n}{12}$
  b) $n + 11$
  c) $n - 8$
4.a) When the Input number increases by 1, the
     Output number increases by 2.
b)

Input x	Output
1	4
2	6
3	8
4	10
5	12
6	14

c) $2x + 2$; The table shows how $2x + 2$ relates to x.

5.a) 3, s, 2
  b) 7, p
  c) 1, c, 8
  d) 11, w, 9
6.a) $5 + 3c$
b)

Additional Half Hours	Cost ($)
0	5
1	8
2	11
3	14
4	17

  c) The graph goes up to the right.
     When the number of additional half hours
     increases by 1, the cost increases by $3.
  d) i) $23.00
     ii) 8 additional half hours
7.a) $x = 5$
  b) $x = 2$
8.a) 11 red tiles
  b) 3 ways: 3 red tiles, or 4 red tiles and 1 yellow
     tile, or 5 red tiles and 2 yellow tiles
9.a) 0
  b) –2
  c) –12
  d) +2

10.a) i) +10, –5
     ii) +25, –10
     iii) –9, +12
  b) i) $(+10) + (-5) = +5$; I deposit $5.
     ii) $(+25) + (-10) = +15$;
        The balloon rises 15 m.
     iii) $(-9) + (+12) = +3$;
        I ride the elevator up 3 floors.
11.a) 115 m or –115 m
  b) –75 m or 75 m
12.a) –4
  b) –6
  c) +10
  d) –6
13.a) i) $0.\overline{03}$
     ii) $0.\overline{06}$
     iii) $0.\overline{09}$
  b) As the numerator of the fraction increases by
     1, the corresponding decimal increases by
     $0.\overline{03}$ each time.
  c) i) $\frac{5}{33}$
     ii) $\frac{8}{33}$
     iii) $\frac{10}{33}$
14.a) From greatest to least:
     $5\frac{1}{3}, 5.3, \frac{21}{4}, 4.9, \frac{24}{5}$
15. 1.873 m
16.a) 7.82
  b) 3.96
  c) 15.17
  d) 4.93
17.a) 21 bottles
  b) 0.375 L
18.a) i) $7.80
     ii) $137.79
  b) i) $1.08
     ii) $19.06

Unit 4 Circles and Area, page 128

4.1 Investigating Circles, page 131
1.a) 12 cm
  b) 16 cm
2.a) 14 cm
  b) 8 cm
3.a) 1.9 cm
  b) 15 cm

4. 0.6 m

5.c) 360°

   d) The sum of the angles at the centre is 360°.

6. 15 glasses; Assumptions may vary. For example: All glasses are cylindrical and they can touch.

7. Answers may vary. For example:
   15 cm, 7.5 cm; 2.5 cm, 1.25 cm; 9.6 cm, 4.8 cm; 8.8 cm, 4.4 cm; 1.5 cm, 0.75 cm; 1.8 cm, 0.9 cm; 2.6 cm, 1.3 cm

8. Answers may vary. For example:
   Fix one end of a measuring tape on the circumference. Walk around the circle with the measuring tape at ground level, until you reach the maximum distance across the circle, which is the diameter. The centre of the circle is the midpoint of the diameter.

4.2 Circumference of a Circle, page 136

1.a) About 31.42 cm   **b)** About 43.98 cm
   c) About 47.12 cm

2.a) About 7.64 cm; about 3.82 cm
   b) About 0.76 m; about 0.38 m
   c) About 12.73 cm; about 6.37 cm

3. Less than; π is greater than 3.

4.a) About 7.5 m
   b) About $33.98, assuming the edging does not have to be bought in whole metres

5.a) The circumference doubles.
   b) The circumference triples.

6. About 71.6 cm

7. No, because π never terminates or repeats. So, the circumference will never be a whole number.

8.a) A dotted line with the marks equally spaced apart
   b) About 289 cm, or 2.89 m
   c) About 346 times

9.a) About 40 075 cm
   b) There would be a gap of about 160 m under the ring. You would be able to crawl, walk, and drive a school bus under the ring.

Unit 4 Mid-Unit Review, page 138

2. Answers may vary, but diameters should be less than 20 cm and greater than 10 cm.

3.a) 3.9 cm **b)** 4.1 cm **c)** 5 cm   **d)** 12.5 cm

4. No, two circles with the same radius are the same (congruent).

5.a) About 37.70 cm   **b)** About 50.27 cm

6.a) i) About 207.35 cm
   ii) About 232.48 cm
   iii) About 188.50 cm

b) The tire has the greatest circumference; it has the greatest diameter, too.

7. About 24.38 m

8.a) About 40.7 cm
   b) About 18.0 cm
   c) About 7.2 cm

9. About 78.54 cm

4.3 Area of a Parallelogram, page 139

1.iii) a) 20 cm²
     b) 9 cm²
     c) 30 cm²

2.a) 312 cm²
   b) 195 mm²
   c) 384 cm²

3.b) The 3 parallelograms have equal areas: 21 cm²

4. Yes; Parallelograms with the same base and height have equal areas.

5.b) 10 cm²

6.a) 5 m   **b)** 3 mm   **c)** 6 cm

7. Answers may vary. For example:
   a) $b = 5$ cm, $h = 2$ cm
   b) $b = 6$ cm, $h = 3$ cm
   c) $b = 7$ cm, $h = 4$ cm

8. The area of the parallelogram is 16 cm². The student may have used the side length, 5 cm, as the height of the parallelogram.

9. No, the areas of Shape A and Shape B are equal.

10.a) 95.04 m²
   b) 132 m²
   c) 36.96 m²; 18.45 m² each

4.4 Area of a Triangle, page 145

2.a) 21 cm²   **b)** 12.5 cm² **c)** 12 cm²
   d) 12 cm²   **e)** 10 cm²   **f)** 8 cm²

3.b) In a right triangle, two heights coincide with the sides.

4.a) 21 cm²
   c) Each parallelogram has area 42 cm².

5.a) 4 cm   **b)** 16 m   **c)** 32 mm

6.b) All triangles in part a have the same area: 6 cm²

7.a) $b = 4$ cm, $h = 7$ cm or $b = 2$ cm, $h = 14$ cm
   b) $b = 10$ cm, $h = 2$ cm or $b = 4$ cm, $h = 5$ cm
   c) $b = 4$ cm, $h = 4$ cm or $b = 2$ cm, $h = 8$ cm

8.a) i) The area doubles.
   ii) The area is 4 times as great.
   iii) The area is 9 times as great.

b) I can triple the base or the height of the triangle.

9.a) 11.7 m^2
 b) About 3 cans of paint
10.a) 17 triangles: 12 small, 4 medium, 1 large
 b) 1 small triangle is $\frac{1}{4}$ of a medium triangle
 and $\frac{1}{16}$ of the large triangle.

 1 medium triangle is $\frac{1}{4}$ of the large triangle
 and 4 times as great as a small triangle.
 The large triangle is 4 times as great as a
 medium triangle and 16 times as great as a
 small triangle.
 c) 12 parallelograms: 9 small, 3 medium
 d) 27.6 cm^2 **e)** 6.9 cm^2
 f) 1.725 cm^2
 g) Small: 3.45 cm^2; medium: 13.8 cm^2
11.a) 92.98 m^2
 b) At least 33 sheets of plywood

4.5 Area of a Circle, page 151
 1.a) About 12.57 cm^2
 b) About 153.94 cm^2
 c) About 153.94 cm^2
 d) About 706.86 cm^2
 2.a) About 28.27 cm^2
 b) About 113.10 cm^2
 c) About 254.47 cm^2
 d) About 452.39 cm^2
 3.a) The area is 4 times as great.
 b) The area is 9 times as great.
 c) The area is 16 times as great.
 4.a) The area of the circle is approximately
 halfway between the area of the smaller
 square and the area of the larger square:
 About 75 cm^2
 b) About 78.54 cm^2
 c) Answers may vary.
 5.a) About 104 cm^2
 b) About 16 cm^2
 6.a) About 0.0707 m^2
 b) About 1.0603 m^2;
 about 3.3929 m^2; about 5.6549 m^2
 7.a) About 113.10 cm^2
 b) About 19.63 cm^2
 c) About 34.58 cm^2
 8. Two large pizzas are the better deal.

4.6 Interpreting Circle Graphs, page 158
 1.a) Traditional dance lessons
 b) Powwow drum classes; traditional dance
 lessons

 c) Stick games: 175 students;
 Powwow drum classes: 200 students;
 traditional dance lessons: 125 students
 2.a) 0 to 12 years and 13 to 19 years
 b) **i)** 112 500 viewers
 ii) 62 500 viewers
 iii) 25 000 viewers
 3.a) 161 t
 b) 805 t
 4.a) French: $550; History: $1050;
 Science: $750; Biography: $550;
 Geography: $450; Fiction: $900;
 Reference: $750
 b) The total amount of money spent on each
 type of book should be $5000.
 5.a) 10%
 b) Saskatchewan, Manitoba, Alberta,
 British Columbia
 c) Saskatchewan: 968 300 people;
 about 968 000 people
 Manitoba: 1 161 960 people;
 about 1 162 000 people
 Alberta: 3 292 220 people;
 about 3 292 000 people
 British Columbia: 4 260 520 people;
 about 4 261 000 people
 6.a) 25 students
 b) Autumn: $\frac{7}{2}$; 28%; winter: $\frac{3}{25}$; 12%;

 spring: $\frac{5}{25}$; 20%; summer: $\frac{10}{25}$; 40%
 c) All percents in part b should add up to 100.
 7.a) Morning Snack Mix: sunflower seeds 30 g,
 almonds 54 g, raisins 25.5 g, peanuts 40.5 g
 Super Snack Mix: raisins 19.5 g, banana
 chips 34.5 g, cranberries 25.5 g, papaya
 chunks 40.5 g, pineapple chunks 30 g
 b) Morning Snack Mix: 51 g of raisins
 Super Snack Mix: 39 g of raisins
 I assumed the percents of the ingredients in
 both snack mixes remain the same.

4.7 Drawing Circle Graphs, page 163
 1.a) 50 students
 b) Blue: $\frac{12}{50} = \frac{6}{25}$; brown: $\frac{24}{50} = \frac{12}{25}$;

 green: $\frac{8}{50} = \frac{4}{25}$; grey: $\frac{6}{50} = \frac{3}{25}$
 c) Blue: 24%; brown: 48%; green: 16%;
 grey: 12%
 2.a) 92 people

b) MAJIC99: $\frac{88}{400} = \frac{11}{50}$, 22%;

EASY2: $\frac{92}{400} = \frac{23}{100}$, 23%;

ROCK1: $\frac{120}{400} = \frac{3}{10}$, 30%;

HITS2: $\frac{100}{400} = \frac{1}{4}$, 25%

3.a) 40 000 000 U.S. residents

b) $\frac{1\,200\,000}{40\,000\,000} = \frac{12}{400} = \frac{3}{100}$

c) 10%

4.a) Yes, each number of students can be written as a fraction of the whole.

b) No, data cannot be written as a fraction of the whole.

5. Asia: about 367 million km^2
Africa: about 244 million km^2
South America: about 147 million km^2
Antarctica: about 98 million km^2
Europe: about 86 million km^2
Australia: about 61 million km^2

Unit 4 Unit Review, page 168

1. Answers may vary. For example: Use a pencil, a string, and a pin.

2.a) 6 cm   **b)** 10 cm   **c)** 3.5 cm

3.a) 30 cm   **b)** 44 cm   **c)** 8.4 cm

4. About 34.85 m

5.a) About 75.40 m **b)** 14 m **c)** About 87.96 m

6.a) About 94.25 mm   **b)** About 131.95 mm

c) Mel's dial; it has the greater radius.

7. Answers may vary. For example: 6 cm and 4 cm; 4 cm and 6 cm; 8 cm and 3 cm; 3 cm and 8 cm; 2 cm and 12 cm; 12 cm and 2 cm; 1 cm and 24 cm; 24 cm and 1 cm

8.a) 3.84 m^2

b) i) 0.96 m^2 **ii)** 13.44 m^2

9.a) Answers may vary. For example: $b = 1$ cm, $h = 24$ cm; $b = 2$ cm, $h = 12$ cm; $b = 3$ cm, $h = 8$ cm; $b = 4$ cm, $h = 6$ cm; $b = 6$ cm, $h = 4$ cm; $b = 8$ cm, $h = 3$ cm; $b = 12$ cm, $h = 2$ cm; $b = 24$ cm, $h = 1$ cm

b) The area of the parallelograms in question 7 is double the area of the triangles in part a.

10. $1265.63

11.a) About 201.06 m^2 **b)** About 50.27 m

12.a) The circumference is halved.

b) The area is one-quarter of what it was.

13. About 637.94 cm^2

14. I calculated the area of each shape: about 55.42 cm^2, 54 cm^2, 56 cm^2
The shape in part c will require the most paint.

15.a) Laura received the most votes.

b) Jarrod: 140 votes; Laura: 280 votes; Jeff: 80 votes

16.a) Lake Huron

b) Lake Superior has the greatest surface area.

c) 26 840 km^2

17.a) Water: 62%, protein: 17%, fat: 15%, nitrogen: 3%, calcium: 2%, other: 1%

b) 37.2 kg

18.a) Manitoba: 10%, Saskatchewan: 10%, Quebec: 30%, Ontario: 50%

Unit 4 Practice Test, page 171

2.a) About 31.42 cm   **b)** About 78.54 cm^2

3. 360°

4.a) 63 cm^2 **b)** 9 cm^2

5.a) Too many to count

b) No, because π never terminates or repeats. So, the area will never be a whole number.

6.b) No. The circle represents the whole and each percent can be written as a fraction of the whole.

Unit 5 Operations with Fractions, page 176

5.1 Using Models to Add Fractions, page 179

1.a) $\frac{2}{4} + \frac{1}{2} = 1$ **b)** $\frac{2}{3} + \frac{4}{6} = 1\frac{1}{3}$ **c)** $\frac{7}{10} + \frac{4}{5} = 1\frac{1}{2}$

2.a) $\frac{7}{8} + \frac{1}{2} = 1\frac{3}{8}$ **b)** $\frac{3}{10} + \frac{2}{5} = \frac{7}{10}$ **c)** $\frac{2}{3} + \frac{1}{2} = 1\frac{1}{6}$

d) $\frac{2}{3} + \frac{5}{6} = 1\frac{1}{2}$ **e)** $\frac{3}{6} + \frac{1}{12} = \frac{7}{12}$ **f)** $\frac{1}{4} + \frac{2}{8} = \frac{1}{2}$

g) $\frac{1}{3} + \frac{1}{2} = \frac{5}{6}$ **h)** $\frac{1}{2} + \frac{4}{10} = \frac{9}{10}$

3. $\frac{1}{2}$ h

4.a) i) $\frac{2}{5}$

ii) 1

iii) $\frac{7}{10}$

iv) $\frac{2}{3}$

b) Answers may vary. For example: Use fraction circles. Or, add numerators.

5.a) $\frac{3}{4}$; less   **b)** $\frac{9}{5} = 1\frac{4}{5}$; greater

c) 1; equal **d)** $\frac{4}{10}=\frac{2}{5}$; less

6. Answers may vary. For example: $\frac{1}{6}$ and $\frac{2}{3}$

7.a) $\frac{1}{8};\frac{1}{4};\frac{3}{8}$

b) $\frac{3}{4};\frac{1}{4}$

5.2 Using Other Models to Add Fractions, page 183

1.a) $\frac{2}{4},\frac{3}{6},\frac{4}{8}$ **b)** $\frac{2}{8}$ **c)** $\frac{4}{6},\frac{6}{9}$

2.a) $\frac{3}{4}+\frac{7}{8}=\frac{13}{8}$ **b)** $\frac{5}{6}+\frac{2}{3}=\frac{9}{6}$ **c)** $\frac{3}{2}+\frac{3}{4}=\frac{9}{4}$

3. Answers may vary. For example:
  a) The greater denominator is a multiple of the lesser denominator. The greater denominator shows which number line to use to get the answer.
  b) One denominator is a multiple of the other.

4.a) $\frac{7}{6}$ **b)** $\frac{11}{12}$ **c)** $\frac{7}{10}$ **d)** $\frac{1}{4}$

5.a) $\frac{5}{6}$ **b)** $\frac{19}{12}$ **c)** $\frac{11}{10}$ **d)** $\frac{13}{15}$

6. Answers may vary. For example:
  a) The least common multiple of the denominators shows which number line to use to get the answer.
  b) The denominators are not multiples, nor factors of each other.
  c) Use a number line divided in fractions whose denominator is given by the least common multiple of the unrelated denominators.

7.a) $\frac{13}{21}$ **b)** $\frac{35}{36}$ **c)** $\frac{57}{40}$ **d)** $\frac{29}{35}$

8. $\frac{19}{12}$

9.a) There are 36 possible fractions:
$$\frac{1}{1},\frac{1}{2},\frac{1}{3},\frac{1}{4},\frac{1}{5},\frac{1}{6},\frac{2}{1},\frac{2}{2},\frac{2}{3},\frac{2}{4},\frac{2}{5},\frac{2}{6},\frac{3}{1},\frac{3}{2},\frac{3}{3},\frac{3}{4},\frac{3}{5},\frac{3}{6},$$
$$\frac{4}{1},\frac{4}{2},\frac{4}{3},\frac{4}{4},\frac{4}{5},\frac{4}{6},\frac{5}{1},\frac{5}{2},\frac{5}{3},\frac{5}{4},\frac{5}{5},\frac{5}{6},\frac{6}{1},\frac{6}{2},\frac{6}{3},\frac{6}{4},\frac{6}{5},\frac{6}{6}$$
Answers may vary.
  For example: $\frac{3}{4}+\frac{5}{6}=\frac{19}{12}=1\frac{7}{12}$; $\frac{3}{4}+\frac{1}{2}=\frac{5}{4}=1\frac{1}{4}$

  b) $\frac{4}{6}+\frac{2}{5}=\frac{16}{15}$

10. Answers may vary. For example:
$$\frac{7}{10}+\frac{4}{5}=\frac{3}{2}; \frac{3}{4}+\frac{3}{4}=\frac{3}{2}$$

11. Yes, $\frac{7}{4}<2$

12. 2 cups

5.3 Using Symbols to Add Fractions, page 188

1.a) Eighths **b)** Twenty-fourths
  c) Ninths **d)** Fifteenths

2.a) 1 **b)** 8 **c)** 2 **d)** 20

3.a) $\frac{7}{9}$ **b)** $\frac{5}{6}$ **c)** $\frac{15}{8}=1\frac{7}{8}$ **d)** $\frac{11}{12}$

4.a) About 1; $\frac{11}{10}=1\frac{1}{10}$ **b)** About $\frac{1}{2}$; $\frac{19}{24}$

  c) About 2; $\frac{29}{18}=1\frac{11}{18}$ **d)** About $1\frac{1}{2}$; $\frac{37}{28}=1\frac{9}{28}$

  e) About $\frac{1}{2}$; $\frac{11}{15}$ **f)** About 1; $\frac{31}{30}=1\frac{1}{30}$

5. $\frac{3}{16}$

6. $\frac{3}{4}+\frac{4}{5}$ is greater.

7. Statement b is true: $\frac{3}{10}+\frac{1}{5}+\frac{1}{2}=1$

  Statement a is false: $\frac{1}{10}+\frac{3}{5}+\frac{1}{2}=\frac{12}{10}=\frac{6}{5}>1$

8. About $\frac{29}{30}$

9. Sums in parts a, e, and f are correct.

10.a) $\frac{13}{8}=1\frac{5}{8}$ **b)** $\frac{43}{20}=2\frac{3}{20}$ **c)** $\frac{35}{18}=1\frac{17}{18}$

Unit 5 Mid-Unit Review, page 190

1. $\frac{3}{5}+\frac{3}{10}=\frac{9}{10}$

2. $\frac{11}{12}$ h

3.a) $\frac{1}{2}+\frac{5}{12}=\frac{11}{12}$ **b)** $\frac{2}{3}+\frac{3}{4}=\frac{17}{12}=1\frac{5}{12}$

4.a) $\frac{5}{8}$ **b)** $\frac{5}{6}$ **c)** $\frac{13}{12}=1\frac{1}{12}$ **d)** $\frac{9}{10}$

5. $\frac{3}{2}=1\frac{1}{2}$; Methods may vary. For example: Use Pattern Blocks. Or, use fraction circles. Or, use equivalent fractions.

6.a) $\frac{9}{8}=1\frac{1}{8}$ **b)** $\frac{14}{15}$ **c)** $\frac{3}{8}$ **d)** $\frac{17}{12}=1\frac{5}{12}$

7. No; $\frac{59}{60}<1$

8. a) i) $\frac{3}{4}$

  ii) $\frac{1}{2}$

iii) $\frac{1}{4}$

iv) $\frac{1}{2}$

b) Puzzles and games

5.4 Using Models to Subtract Fractions, page 193

1. Answers may vary. For example:

a) $\frac{4}{8}$ and $\frac{5}{8}$ **b)** $\frac{3}{12}$ and $\frac{4}{12}$

c) $\frac{4}{6}$ and $\frac{1}{6}$ **d)** $\frac{6}{10}$ and $\frac{5}{10}$

2.a) $\frac{1}{3}$; Less than $\frac{1}{2}$ **b)** $\frac{3}{4}$; Greater than $\frac{1}{2}$

c) $\frac{1}{3}$; Less than $\frac{1}{2}$ **d)** $\frac{1}{6}$; Less than $\frac{1}{2}$

3.a) $\frac{1}{4}$ **b)** $\frac{3}{5}$ **c)** $\frac{1}{3}$ **d)** $\frac{1}{4}$

4.a) Subtract the numerators only. The denominator remains the same.

    b) Examples may vary.

5.a) $\frac{7}{9} - \frac{1}{3} = \frac{4}{9}$ **b)** $\frac{7}{8} - \frac{3}{4} = \frac{1}{8}$

  c) $\frac{8}{10} - \frac{2}{5} = \frac{4}{10} = \frac{2}{5}$ **d)** $\frac{11}{12} - \frac{2}{3} = \frac{3}{12}$

6.a) $\frac{1}{8}$ **b)** $\frac{1}{5}$ **c)** $\frac{3}{8}$ **d)** $\frac{7}{12}$

7. $\frac{1}{6}$

8. $\frac{1}{4}$

9. No. Spencer needs $\frac{1}{12}$ cup more.

10. Answers may vary. For example:

    a) $\frac{2}{3} - \frac{1}{3} = \frac{1}{3}$ **b)** $\frac{4}{5} - \frac{1}{5} = \frac{3}{5}$ **c)** $\frac{2}{3} - \frac{2}{4} = \frac{1}{6}$

11.a) More: $\frac{3}{4} - \frac{1}{8} = \frac{5}{8} > \frac{1}{2}$ **b)** $\frac{1}{8}$

12.a) iii   **b)** Use estimation.

5.5 Using Symbols to Subtract Fractions, page 197

1.a) $\frac{2}{5}$ **b)** $\frac{1}{3}$ **c)** $\frac{1}{3}$ **d)** $\frac{2}{7}$

2.a) $\frac{1}{2}$ **b)** $\frac{1}{8}$ **c)** $\frac{4}{5}$ **d)** $\frac{1}{12}$

3.a) $\frac{1}{12}$ **b)** $\frac{2}{15}$ **c)** $\frac{19}{20}$ **d)** $\frac{1}{10}$

4.a) $\frac{1}{6}$ **b)** $\frac{11}{12}$ **c)** $\frac{17}{30}$ **d)** $\frac{1}{12}$

5. Walnuts; $\frac{1}{12}$ cup more

6.a) Terri; $1\frac{5}{12} > 1\frac{1}{4}$

    b) $\frac{1}{6}$ h

7. Answers may vary. For example: $\frac{9}{4} - \frac{3}{2} = \frac{3}{4}$

8. The other fraction is between $\frac{1}{2}$ and $\frac{3}{4}$.

9. 18 min

5.6 Adding with Mixed Numbers, page 202

1.a) $\frac{3}{2}$ **b)** $\frac{17}{4}$ **c)** $\frac{7}{4}$ **d)** $\frac{18}{5}$

2.a) $3\frac{2}{5}$ **b)** $2\frac{1}{4}$ **c)** $4\frac{1}{2}$ **d)** $4\frac{2}{3}$

3.a) $1\frac{1}{2}$ **b)** $2\frac{1}{3}$ **c)** $4\frac{1}{6}$ **d)** $6\frac{1}{6}$

4.a) 6 **b)** $4\frac{3}{4}$ **c)** $7\frac{7}{9}$ **d)** $8\frac{2}{5}$

5.a) $3\frac{3}{8}$ **b)** $3\frac{1}{12}$ **c)** $5\frac{1}{8}$ **d)** $4\frac{1}{10}$

6.a) $3\frac{7}{10}$ **b)** $2\frac{7}{10}$ **c)** $5\frac{7}{10}$ **d)** $7\frac{7}{10}$

7.a) $3\frac{7}{12}$ **b)** $2\frac{2}{5}$ **c)** $3\frac{7}{20}$ **d)** $2\frac{13}{14}$

    e) $6\frac{13}{24}$ **f)** $5\frac{4}{15}$ **g)** $7\frac{11}{40}$ **h)** $6\frac{1}{12}$

8. $6\frac{7}{15}$ h

9.a) Estimates may vary. For example: About $3\frac{1}{2}$

    b) $3\frac{5}{8}$

10. $9\frac{5}{12}$ cups

11.a) $3\frac{7}{10}$

    b) $\frac{8}{5}$ and $\frac{21}{10}$

    c) $\frac{37}{10}$

12. $4\frac{5}{12}$ h

13. $1\frac{2}{5}$ or $\frac{7}{5}$; equivalent fractions may vary.

5.7 Subtracting with Mixed Numbers, page 207

1.a) $1\frac{1}{5}$ **b)** $2\frac{1}{4}$ **c)** 3 **d)** $\frac{5}{3} = 1\frac{2}{3}$

2.a) $1\frac{1}{3}$ **b)** 2 **c)** $\frac{1}{2}$ **d)** $1\frac{3}{4}$

3.a) $2\frac{1}{6}$ **b)** $1\frac{1}{6}$ **c)** $2\frac{1}{6}$ **d)** $4\frac{1}{6}$

4.a) About $2\frac{1}{2}$; $\frac{9}{4} = 2\frac{1}{4}$

b) About $1\frac{1}{2}$; $\frac{3}{2} = 1\frac{1}{2}$

c) About $\frac{1}{2}$; $\frac{13}{20}$

d) About $1\frac{1}{2}$; $\frac{13}{20} = 1\frac{3}{10}$

5.a) i) $\frac{11}{5} = 2\frac{1}{5}$　　ii) $\frac{25}{7} = 3\frac{4}{7}$

　　iii) $\frac{25}{6} = 4\frac{1}{6}$　　iv) $\frac{50}{9} = 5\frac{5}{9}$

6.a) $2\frac{11}{20}$ **b)** $1\frac{2}{5}$ **c)** $2\frac{5}{12}$ **d)** $2\frac{1}{21}$

7.i) **a)** $2\frac{3}{10}$ 　　　　**b)** $\frac{23}{10}$

c) Answers may vary. For example:

The first method is easier because $\frac{3}{5}$ is

greater than $\frac{3}{10}$.

ii) **a)** $1\frac{7}{10}$ 　　　　**b)** $\frac{17}{10}$

c) Answers may vary. For example:

The second method is easier because $\frac{3}{5}$

is less than $\frac{3}{10}$.

8. $1\frac{17}{40}$ cups

9. $\frac{11}{12}$ h

10.a) $\frac{19}{24}$ 　　　　**b)** $\frac{31}{18}$ or $1\frac{13}{18}$

c) $\frac{44}{15}$ or $2\frac{14}{15}$ **d)** $\frac{101}{40}$ or $2\frac{21}{40}$

11.a) Estimates may vary.

For example: About $1\frac{1}{2}$

b) $\frac{35}{24}$ or $1\frac{11}{24}$ 　**d)** $\frac{29}{24}$ or $1\frac{5}{24}$

12. Answers may vary. For example: $\frac{21}{8}$ or $2\frac{5}{8}$

Unit 5 Unit Review, page 213

1.a) $\frac{13}{12}$ **b)** 1 **c)** $\frac{11}{12}$ **d)** $\frac{7}{10}$

2.a) $\frac{11}{9}$ **b)** $\frac{3}{2}$ **c)** $\frac{3}{4}$ **d)** $\frac{9}{8}$

3. Answers may vary. For example: $\frac{1}{4} + \frac{3}{8} = \frac{5}{8}$

4. Answers may vary. For example:

a) $\frac{12}{20}$ and $\frac{15}{20}$ 　　**b)** $\frac{2}{5}$ and $\frac{1}{5}$

c) $\frac{8}{18}$ and $\frac{9}{18}$ 　　**d)** $\frac{15}{24}$ and $\frac{4}{24}$

5.a) $\frac{4}{5}$ **b)** $\frac{13}{14}$ **c)** $\frac{29}{30}$ **d)** $\frac{17}{20}$

6.a) $1 - \frac{1}{3} = \frac{4}{6}$ 　　**b)** $\frac{7}{10} - \frac{2}{5} = \frac{3}{10}$

c) $\frac{10}{12} - \frac{3}{4} = \frac{1}{12}$ 　**d)** $\frac{5}{8} - \frac{1}{4} = \frac{3}{8}$

7.a) $\frac{3}{5}$ 　　**b)** $\frac{1}{2}$ 　　**c)** $\frac{5}{12}$

8.a) Javon; $\frac{5}{6} > \frac{7}{9}$ **b)** $\frac{1}{18}$

9.a) $\frac{1}{2}$ 　　　　**b)** $\frac{3}{2} = 1\frac{1}{2}$

c) $\frac{27}{20} = 1\frac{7}{20}$ 　**d)** $\frac{19}{12} = 1\frac{7}{12}$

10. Answers will vary. For example:

a) $\frac{4}{3} - \frac{5}{6} = \frac{1}{2}$ **b)** $\frac{31}{36} - \frac{1}{9} = \frac{3}{4}$ **c)** $\frac{17}{20} - \frac{3}{4} = \frac{1}{10}$

d) $\frac{5}{2} - \frac{7}{3} = \frac{1}{6}$ **e)** $\frac{5}{6} - \frac{7}{12} = \frac{1}{4}$

11.a) Brad 　　　**b)** $\frac{1}{8}$ bottle

12. $\frac{3}{8}$

13.a) $6\frac{2}{3}$ **b)** $1\frac{7}{12}$ **c)** $5\frac{1}{2}$ **d)** $6\frac{13}{20}$

14.a) $4\frac{1}{2}$ **b)** $4\frac{5}{8}$ **c)** $10\frac{1}{10}$ **d)** $8\frac{2}{9}$

15. $3\frac{5}{8}$ h

16.a) $\frac{33}{8}$, or $4\frac{1}{8}$ 　　**b)** $\frac{25}{9}$, or $2\frac{7}{9}$

c) $\frac{19}{12}$, or $1\frac{7}{12}$ 　　**d)** $\frac{47}{24}$, or $1\frac{23}{24}$

17.a) The second recipe; $1\frac{7}{9} > 1\frac{3}{4}$

b) $\frac{1}{8}$ cup

18.a) $\frac{25}{6}$, or $4\frac{1}{6}$ 　　**b)** $\frac{49}{30}$, or $1\frac{19}{30}$

c) $\frac{169}{24}$, or $7\frac{1}{24}$ 　**d)** $\frac{3}{4}$

19. $\frac{5}{6}$ h

Unit 5 Practice Test, page 215

1.a) 2 　　　**b)** $\frac{19}{30}$

c) $\frac{1}{4}$ **d)** $\frac{29}{18} = 1\frac{11}{18}$

2. Answers may vary. For example:

a) $\frac{1}{5} + \frac{2}{5} = \frac{3}{5}$ **b)** $\frac{1}{35} + \frac{4}{7} = \frac{3}{5}$

3. Answers may vary. For example:

a) $\frac{3}{8} - \frac{1}{8} = \frac{1}{4}$ **b)** $\frac{3}{4} - \frac{1}{2} = \frac{1}{4}$

4.a) $\frac{343}{40}$, or $8\frac{23}{40}$ **b)** $\frac{13}{10}$, or $1\frac{3}{10}$

5. $7\frac{3}{4}$ h; Answers may vary. For example: No, Lana cannot do all the jobs. If she allows at least 3 h to travel from one place to another and $\frac{1}{2}$ h for her lunch break, her total time is $11\frac{1}{4}$ h.

6.a) $\frac{1}{2} + \frac{1}{4} = \frac{3}{4}$ **b)** $\frac{1}{2} + \frac{1}{8} = \frac{5}{8}$

7. Answers may vary. For example:

Counter 1: $\frac{1}{6}$ and $\frac{7}{12}$, Counter 2: $\frac{5}{12}$ and $\frac{2}{3}$

Unit 6 Equations, page 218

6.1 Solving Equations, page 223

1.a) equation **b)** expression **c)** expression
d) equation **e)** expression **f)** equation
2.a) $w = 12$ **d)** $x = 96$ **f)** $z = 11$
3.a) $x - 10 = 35$ **b)** $x = 45$
4.a) $7 + n = 18; n = 11$ **b)** $n - 6 = 24; n = 30$

c) $5n = 45; n = 9$ **d)** $\frac{n}{6} = 7; n = 42$

e) $4n + 3 = 19; n = 4$
5.a) $14x = 182; x = 13$ **b)** $b - 14 = 53; b = 67$
c) $100 = 56 + 11p; p = 4$
6. For example: **a)** $4s = 48$ **b)** $s = 12$

7. For example: **a)** $\frac{p}{6} = 11$ **b)** $p = 66$

8. Answers may vary. For example:
a) The perimeter of a triangle is 27 cm. Write an equation you can solve to find the side length of the triangle.
b) $27 = 3t$ **c)** $t = 9$
9.a) $130 = 10 + 24f$ **b)** $f = 5$
10.a) $n = 9$ **b)** $n = 12$ **c)** $n = 15$ **d)** $n = 81$
11.a) $x = 3$ **b)** $y = 6$ **c)** $z = 2166$ **d)** $x = 5$

6.2 Using a Model to Solve Equations, page 229

1.a) $A = 30$ g **b)** $B = 65$ g
c) $C = 50$ g **d)** $D = 21$ g

2.b) **i)** $x = 7$
ii) $x = 14$
iii) $y = 3$
iv) $m = 7$
v) $k = 8$
vi) $p = 21$
3. i) **a)** $5 + n = 24$ **b)** $n = 19$
ii) **a)** $n + 8 = 32$ **b)** $n = 24$
iii) **a)** $3n = 42$ **b)** $n = 14$
iv) **a)** $2n + 5 = 37$ **b)** $n = 16$
4.a) $60 = 12h; h = 5$ m **b)** $112 = 8h; h = 14$ cm
c) $169 = 13h; h = 13$ m
5.a) Left pan: x and 35 g; right pan: 35 g and 25 g
b) $x = 25$
6. Problems may vary. For example:
a) Helen is 16 years old. Kian is 4 years younger than Helen. How old is Kian?
b) Helen is 4 years older than Kian. Kian is 16 years old. How old is Helen?
c) Part a: $x = 12$; part b: $x = 20$
7. Answers may vary. The sum of the digits should be a multiple of nine. For example:
$5 + x + 7 = 18, x = 6$;
567 is divisible by 9.

6.3 Solving Equations Involving Integers, page 234

1.a) $x = 4$ **b)** $x = 7$ **c)** $x = 10$
d) $x = 12$ **e)** $x = 13$ **f)** $x = 14$
2.a) $n = 13$ **b)** $x = 2$ **c)** $p = 7$
d) $x = -5$ **e)** $s = -14$ **f)** $x = 3$
3. $x = 17$
4. $f - 6 = 5; f = 11$
5.a) $t - 8 = -3$ **b)** $t = 5$
6.a) $x = 7$ **b)** $n = 19$
7.a) $n + 2 = 4; +2$
b) $n - 2 = 1; +3$
c) $n - 4 = -2; +2$

Unit 6 Mid-Unit Review, page 236

1.a) **i)** $5 + d = 12; d = 7$
ii) $2d = 12; d = 6$
b) **i)** $67 + s = 92; s = 25$
ii) $3w + 8 = 29; w = 7$
2. i) **a)** $n + 9 = 17$ **c)** $n = 8$
ii) **a)** $3n = 21$ **c)** $n = 7$
iii) **a)** $7 + 2n = 19$ **c)** $n = 6$
3. $40 = 14 + 2B$; Bill is 13 years old.
4. i) **a)** $n - 8 = 7$ **c)** $n = 15$
ii) **a)** $t - 6 = -4$ **c)** $t = 2$
iii) **a)** $m - 7 = 5$ **c)** $m = 12$

6.4 Solving Equations Using Algebra, page 238

1.a) $x = 62$ **b)** $x = 12$ **c)** $x = 17$
2.a) $19 + n = 42$; $n = 23$
  b) $3n + 10 = 25$; $n = 5$
  c) $15 + 4n = 63$; $n = 12$
3.a) $27 = 5 + 2J$ **b)** $J = 11$
4.a) $33 = 3 + 6h$ **b)** $h = 5$
5.a) $25 = 4 + 7x$ **b)** $x = 3$
6.a) $56 = 24 + 4s$ **b)** $s = 8$
7.a) $72 + 24w = 288$ **b)** $w = 9$; After 9 weeks
8. Problems may vary. For example:
  a) Sarah spent $9 at the bowling alley. How many games did she bowl?
  b) $9 = 3 + 2g$; $g = 3$
9.a) 17   **b)** 13   **c)** 27

6.5 Using Different Methods to Solve Equations, page 243

1.a) $x = 8$ **b)** $x = 21$ **c)** $x = 64$ **d)** $x = 50$
2. Methods may vary.
  a) $x = 7$ **b)** $x = 17$
  c) $x = 54$ **d)** $x = -13$
  e) $x = 9$ **f)** $x = 7$
  g) $x = 7$ **h)** $x = 11$
3.a) $x + 7 = 21$; $x = 14$
4. $\frac{c}{8} = 4$; $c = 32$
6.a) For example: $20 + 8m = 92$; $m = 9$
  b) Methods may vary. For example: I used algebra.
7.a) $37 = 5 + 4g$; $g = 8$ **b)** $37 = 10 + 9g$; $g = 3$
8.a) $85 = 40 + 15n$; $n = 3$
  b) $140 = 90 + 10n$; $n = 5$
9.b) Answers may vary. For example:
  $15 + 8 + 12 = 35$ or $25 + 8 + 2 = 35$

Unit 6 Reading and Writing in Math: Decoding Word Problems, page 247

1. One group of 6 rows by 6 columns; 4 groups of 3 rows by 3 columns; 9 groups of 2 rows by 2 columns
2. 144 fence posts
3. 12:21, 1:01, 1:11, 1:21, 1:31, 1:41, 1:51, 2:02, 2:12, 2:22, 2:32, 2:42, 2:52, 3:03, 3:13, 3:23, 3:33, 3:43, 3:53, 4:04, 4:14, 4:24, 4:34, 4:44, 4:54, 5:05, 5:15, 5:25, 5:35, 5:45, 5:55, 6:06, 6:16, 6:26, 6:36, 6:46, 6:56, 7:07, 7:17, 7:27, 7:37, 7:47, 7:57, 8:08, 8:18, 8:28, 8:38, 8:48, 8:58, 9:09, 9:19, 9:29, 9:39, 9:49, 9:59, 10:01, 11:11

Unit 6 Unit Review, page 248

1. $x = 13$; Jan started with 13 stamps.
2.a) $5 + n = 22$; $n = 17$ **b)** $n - 7 = 31$; $n = 38$
  c) $6n = 54$; $n = 9$ **d)** $\frac{n}{8} = 9$; $n = 72$
  e) $9 + 3n = 24$; $n = 5$
3.a) $m - 36 = 45$; $m = 81$ **b)** $13b = 208$; $b = 16$
  c) $\frac{d}{15} = 17$; $d = 255$
4.a) $27 = 15 + x$; $x = 12$
  b) $25 = 2x + 11$; $x = 7$
5.a) $x = 6$ cm   **b)** $x = 16$ cm
6.a) $81 = 25 + 8c$; $c = 7$
7.a) $x = 3$ **b)** $n = -3$ **c)** $w = 15$ **d)** $x = 15$
8.a) $5 + x = -7$, $y - 5 = 7$
  b) $x = -12$, $y = 12$
9.i) **a)** $-8 + x = 3$ **b)** $x = 11$
  ii) **a)** $3 + y = -1$ **b)** $y = -4$
10.a) $56 = 7n$ **b)** $n = 8$
11.a) $400 = 140 + x$ **b)** $x = 260$
12.a) $228 = 4p$ **b)** $p = 57$
13.a) $x = 19$ **b)** $x = 7$ **c)** $x = 45$ **d)** $x = 8$
14.a) $x = 12$ **b)** $x = -10$ **c)** $x = 3$
  d) $x = 7$ **e)** $x = 99$ **f)** $x = 13$
15. $25 = 1 + 3b$; $b = 8$
16.a) $545 = 125 + 12m$ **b)** $m = 35$

Unit 6 Practice Test, page 251

1.a) $x = 2$ **b)** $p = 14$
  c) $c = 63$ **d)** $q = 13$
2.a) $44 = 4h$; $h = 11$
  b) $50 = 2b + 32$; $b = 9$
3.a) 10 km
  b) 48 km
  c) 58 km
4.a) $47 = 12 + 5d$; $d = 7$ **b)** $107 = 12 + 5d$; $d = 19$
5.a) $75 + 3 \times 25$ **b)** $204 = 75 + 3s$; $s = 43$

Cumulative Review Units 1–6, page 254

1.a) 1, 2, 3, 4, 5, 6, 8, 10, 12, 15, 20, 24, 30, 40, 60, 120
  b) 1, 2, 3, 4, 6, 7, 12, 14, 21, 28, 42, 84
  c) 1, 2, 3, 4, 6, 8, 9, 12, 18, 24, 27, 36, 54, 72, 108, 216
2.a) $x + 7 = 19$
  b) $x = 12$
3.a) -8 **b)** -10 **c)** $+9$
4.a) -6 **b)** $+12$
  c) $+6$ **d)** -12
5. Answers may vary. For example:
  a) $1.\overline{6}$ **b)** 0.6

c) 2.2   **d)** 2.75
6. 56.16 m^2
7.a) $71.99
  b) $82.07
8.a) Too many to count
  b) Too many to count
9.a) About 37.7 cm
10. Greatest area: part b; least area: part c
11.a) 50 m
  b) About 7.96 m
  c) About 199.06 m^2
12.a) 120 students

  b) Black: $\frac{60}{120} = \frac{1}{2}$; brown: $\frac{20}{120} = \frac{1}{6}$;

  blonde: $\frac{30}{120} = \frac{1}{4}$; red: $\frac{10}{120} = \frac{1}{12}$

  c) Black: 50%; brown: about 17%;
  blonde: 25%; red: about 8%

13. $\frac{17}{24}$ cup of sugar

14.a) $\frac{23}{30}$ **b)** $\frac{5}{12}$

  c) $\frac{13}{24}$ **d)** $\frac{17}{36}$

15. $\frac{5}{8}$

16.a) $8\frac{11}{12}$ **b)** $\frac{19}{30}$

  c) $5\frac{4}{15}$ **d)** $1\frac{5}{24}$

17.a) **i)** $s = 5$
     ii) $s = 9$
     iii) $s = 9$
     iv) $s = 6$
18.a) $x = 6$ **b)** $x = 17$
19.a) $7x + 5 = 250$
  b) $x = 35$; Juan worked 35 h.
20.a) $x + 3 = 10; x = 7$
    Shin's score after Round One was +7.
  b) $x - 1 = -4; x = -3$
    Lucia's score after Round One was –3.

Unit 7 Data Analysis, page 256

7.1 Mean and Mode, page 260
  1.a) 4
    b) 3
    c) 3
  2.a) 6
    b) 34
  3.a) 4
    b) no mode

4.a) $13
  b) $15
  c) The mean is $14.50. The mode remains
    the same: $15
5.a) Mean: 29.5; mode: 18
  b) Answers will vary. For example: 10, 13, 15,
    15, 21, 28, 36, 36, 45, 54, 60
6.

		Mean	Mode
a)	Games Played	55	no mode
b)	Goals	23.25	no mode
c)	Assists	29	39
d)	Points	52.25	no mode

7.a) Volleyball and soccer
  b) I could count the number of bars of
    equal length.
    The length which occurs most often is the
    mode. Mode: 750 people
  c) About 1003
8.a) Any pair of numbers whose sum is 11: 0 and
    11, 1 and 10, 2 and 9, 3 and 8, 4 and 7,
    5 and 6
  b) 3 and 8

7.2 Median and Range, page 264
  1.a) Median: 90; range: 20
    b) Median: 25.5 kg; range: 73 kg
  2.a) Class A: 12.5; Class B: 12
    b) Class A: 7; Class B: 4
    c) Class A; Class A's median mark is greater.
  3.a) **i)** Mean: 7; median: 7; no mode
      ii) Mean: 60; median: 60; modes: 50, 70
      iii) Mean: 56; median: 68; mode: 71
      iv) Mean: 13; median: 13; mode: 13
    b) i, ii, and iv; iv; iii
  4. Answers may vary. For example:
    a) 85, 90, 100, 100, 110, 115
    b) 80, 85, 100, 100, 105, 110
  5. Answers may vary. For example (in cm):
    a) 135, 143, 146, 155, 158, 158, 160, 163, 164,
      166
    b) 150, 154, 158, 163, 163, 163, 165, 170, 174,
      178
  6.a) Median: 120 s; mode: 118 s
    b) 122 s
    c) The mean would be most affected.
      The mean increases to 135.7 s.
      The mode remains 118 s.
      The median increases to 122 s.

7. a)

	Mean	Median	Mode
Games	12.4	12	11
Goals	9.7	6	3
Assists	10.9	10.5	4
Points	20.7	17	10
Penalty Minutes	18.3	8	2 and 8

8. Edward's answer is correct.

7.3 The Effects of Outliers on Average, page 269

1. a) Mean: 4.96 min; median: 5 min; mode: 5 min
   b) The outliers are 0, 1, 2.
   c) Mean: about 5.8 min; median: 5 min; mode: 5 min
   The mean increases. The median and the mode remain the same.
2. a) Mean: 21.35 min; median: 18 min; mode: 15 min
   b) The outlier is 95 min.
   Explanations may vary.
   c) Mean: about 17.47 min; median: 18 min; mode: 15 min
   The mean decreases. The median and the mode remain the same.
   d) About 18 min; Bryan should use the median time to answer.
3. a) Mean: 34.4; median: 36; mode: 36
   b) The outlier is 4.
   c) Mean: about 36.6; median: 36; mode: 36
   The mean increases. The median and the mode remain the same.
   d) No. The outlier is a recording error.
4. a) Mean: about 67.6; median: 68; modes: 65 and 68
   b) The outlier is 0.
   c) Mean: about 73.7; median: 68; modes: 65 and 68
   The mean increases. The median and the modes remain the same.
   d) No. The outlier does not represent the data.
5. a) Examples will vary. For example: The outliers should be ignored when reporting pulse rates.
   b) Examples will vary. For example: The outliers cannot be ignored when reporting average daily temperatures.
6. a) 460 raisins
   b)   **i)** Mean: About 454.5; median: 465; no mode
   ii) 400 and 499

iii) Mean: About 455.2; median: 465; no mode; The mean increases. The median and the mode remain the same.
iv) Yes. These outliers provide important information.
v) No. The mean is significantly less than 460.
7. a) Mean: 5; median: 5; mode: 5 **b)** 19

7.4 Applications of Averages, page 273

1. a) Mean: About 26.4°C; median: 27°C; modes: 23°C and 28°C
   b) The mean best describes the daily high temperature.
   c) No. Explanations may vary. For example: The weather channel reported one of the mode temperatures. The mean and the median are significantly higher than 23°C.
2. a) Math: Mean: About 74.6; median: 75; no mode
   Music: Mean: About 77.3; median: 81; mode: 81
   French: Mean: About 74.4; median: 74; mode: 74
   b) The mean is not one of Caitlin's marks. The median is the middle value in each ordered set of marks. The mode represents the mark that occurs most often.
   c) Caitlin is best at music because the mean, median, and mode are highest for this subject. Caitlin is worst at French because the mean, median, and mode are lowest for this subject.
3. a) Week 1: Mean: $825; median: $800; no mode
   Week 2: Mean: $825; median: $775; no mode
   b) Mean: $825; median: $787.50; mode: $600
   c) The means are the same. The medians and modes are different.
   d) The median best represents the tips earned.
4. a) Mean: About $62 667; median: $50 000; modes: $50 000 and $28 000
   b) $102 000
   c) i) Mean   **ii)** The lesser mode
5. a) Yes   **b)** No
6. a) Mode   **b)** Mean   **c)** Median
7. a) Mean: About 395.3 g; median: 395 g; mode: 405 g
   b) 25 g   **c)** Mode
8. a) i) 85%   **ii)** 90%   **iii)** 95%

b) No, Andrew cannot get a mean mark of 84%
or higher because he would need a math
mark greater than 100%.

9. No, Celia's reasoning is not correct. Her mean
mark is 83.5%.

Technology: Using Spreadsheets to Investigate Averages, page 277

1.a) Mean: About $15.68; median: $15; mode: $9

2.a) Mean: About $51.23; median: $47.19;
mode: $34.45

3. Mean: 110.9; median: 113; no mode

Unit 7 Mid-Unit Review, page 278

1.a) Mean: 165 cm; median: 166 cm;
mode: 170 cm

b) 20 cm

2. Answers may vary. For example: 13, 15, 23,
24, 25; 5, 17, 23, 25, 30

3.a) Mean: About $82.13; median: $75;
mode: $75

b) The outlier, $20, may be a recording error.
The outlier, $229, may be the rate charged
for a luxury suite.

c) Mean: About $76.07; median: $75;
mode: $75
The mean decreases. The median and the
mode remain the same.

d) The outlier, $20, is a recording error and
should not be used. The outlier, $229, is an
actual rate and should be used.

4.a) Mean: About 99.8 g; median: About 99.8 g;
mode: 100.3 g

b) Mode

5.b) False

7.5 Different Ways to Express Probability, page 282

1.a) $\frac{1}{3}$, or about 33.3%, or 1:3

b) 0, or 0%

c) $\frac{2}{16}$, or $\frac{1}{8}$, or 12.5%, or 1:8

b) 1, or $\frac{100}{100}$, or 100%, or 1:1

2.a) $\frac{14}{54}$, or $\frac{7}{27}$, or about 26%, or 7:27

b) $\frac{12}{54}$, or $\frac{2}{9}$, or about 22%, or 2:9

3.a) $\frac{1}{250}$, or 0.4%, or 1:250

b) $\frac{10}{250}$, or $\frac{1}{25}$, or 4%, or 1:25

c) $\frac{225}{250}$, or $\frac{9}{10}$, or 90%, or 9:10

4.a) $\frac{5}{20}$, or $\frac{1}{4}$, or 25%, or 1:4

b) $\frac{11}{20}$, or 55%, or 11:20

c) 1, or 100%, or 1:1

d) 0, or 0%, or 0:20

e) $\frac{1}{20}$, or 5%, or 1:20

5.a) $\frac{1}{8}$, or 12.5%, or 1:8

b) $\frac{7}{8}$, or 87.5%, or 7:8

c) $\frac{4}{8}$, or $\frac{1}{2}$, or 50%, or 1:2

d) $\frac{4}{8}$, or $\frac{1}{2}$, or 50%, or 1:2

e) 0, or 0%, or 0:8   **f)** 1, 100%, 1:1

6. Answers may vary. For example:
You roll a die.

a) The probability of getting a number
less than 10

b) The probability of getting an even number

c) The probability of getting a 4

d) The probability of getting a 7

7. I divided the spinner into 10 equal sectors:
2 red, 5 yellow, 1 blue, and 2 green

8.a) The third candy is most likely white.

b) $\frac{3}{7}$, or about 43%, or 3:7

c) $\frac{4}{7}$, or about 57%, or 4:7

7.6 Tree Diagrams, page 287

1.a) 3H, 3T, 4H, 4T, 5H, 5T, 6H, 6T,
7H, 7T, 8H, 8T
The outcome of rolling a die does not depend
on the outcome of tossing a coin.

b) 1B, 1Y, 1P, 2B, 2Y, 2P, 3B, 3Y,
3P, 4B, 4Y, 4P
The outcome of rolling a tetrahedron does
not depend on the outcome of spinning the
pointer on a spinner.

c) 1, 1; 1, 2; 1, 3; 1, 4; 1, 5; 1, 6; 2, 1; 2, 2; 2, 3;
2, 4; 2, 5; 2, 6; 3, 1; 3, 2; 3, 3; 3, 4; 3, 5; 3, 6;
4, 1; 4, 2; 4, 3; 4, 4; 4, 5; 4, 6; 5, 1; 5, 2; 5, 3;
5, 4; 5, 5; 5, 6; 6, 1; 6, 2; 6, 3; 6, 4; 6, 5; 6, 6
The outcome of rolling one die does not
depend on the outcome of rolling the other
die.

2. Aseea; $\frac{3}{4}$ is greater than $\frac{1}{3}$.

3. Answers may vary. For example:
The probability of rolling an even number

4. The probability of rolling both numbers greater

than 4 is: $\frac{4}{36}$, or $\frac{1}{9}$

5. a)

Paint Colour

Seat Colour		Black	Blue	Red	Silver	Gold
	Grey	Gr, Bla	Gr, Blu	Gr, R	Gr, S	Gr, G
	Black	Bla, Bla	Bla, Blu	Bla, R	Bla, S	Bla, Go

b) $\frac{2}{10}$, or $\frac{1}{5}$, or 20%

6. The player should choose to roll the tetrahedron twice to have the greatest probability of winning.

Unit 7 Unit Review, page 292

1.a) Under par: 10; at par: 2; over par: 7
  b) 26
  c) Mean: About 34.3; median: 35; mode: 33
2. Answers will vary.

For example: 4, 5, $6\frac{1}{2}$, $6\frac{1}{2}$, 7, $7\frac{1}{2}$, 8,

or 4, 5, 5, 6, 7, 8, 8, 9

3.a) Mean: 12.6 h; median: 13.5 h; mode: 15 h
  b) 3 h
  c) Mean: About 13.7 h; median: 15 h; mode: 15 h
  The mean and median decrease. The mode remains the same.
  d) No. The outlier is not typical of the number of hours Josephine works in a week.
4.a) Mean: About 46.3 min; median: 40.5 min; mode: 47 min
  b) 8 min, 74 min, 125 min
  Mean: About 40.1 min; the mean decreases. So, it is greatly affected by the outliers.
  c) Median
  d) Yes, the outliers are actual times spent by students on math homework.
5.a) Mean: 122 s; median: 119.5 s; mode: 118 s
  b) Median   **c)** 19 s
  d) Annette must get a time greater than or equal to 120 s in her next run.
  e) 113 s; unlikely
6.a) Mode   **b)** Median **c)** Mean   **d)** Median

7.a) $\frac{10}{20}$, or $\frac{1}{2}$, or 50%, or 1:2

  b) $\frac{5}{20}$, or $\frac{1}{4}$, or 25%, or 1:4

  c) $\frac{8}{20}$, or $\frac{2}{5}$, or 40%, or 2:5

  d) 0, or 0%, or 0:20   **e)** 1, or 100%, or 1:1
8.a) 2, 3, 4, 6, 8, 9, 12
  b) The probability of getting a product of 2, 3, 8, 9, and 12: $\frac{1}{9}$

  The probability of getting a product of 4 and 6: $\frac{2}{9}$

  c) 2, 3, 8, 9, and 12; 4 and 6
  d) $\frac{8}{9}$

9.b) i) $\frac{1}{3}$ **ii)** $\frac{1}{3}$

   iii) $\frac{1}{9}$ **iv)** $\frac{1}{9}$

11. No, each player has a 50% probability of winning and each prize has a greater value than the cost.

Unit 7 Practice Test, page 295

1.a) 243.25 s   **b)** 208 s
  c) 158 s   **d)** 237.5 s
2.a) Mean: about 7.8; median: 7.25; mode: 7
  b) 18
  c) Mean: about 7.3; median: 7; mode: 7
  The mean and the median decrease. The mode remains the same.
  d) No. The outlier is a recording error.
3.a) ii)   **b)** i   **c)** iv   **d)** iii

Unit 8 Geometry, page 298

8.1 Parallel Lines, page 302
1. Parts a and c
4. Answers may vary. For example:
Use tracing paper.
5. Answers may vary. For example:
Shelves on a bookshelf
6. JE and AB, CL and BK, BE and AF, BF and GK, AF and GK

8.2 Perpendicular Lines, page 305
1. Parts a and b
4. Answers may vary. For example: Book covers, desks, floor, ceiling
5. AE and FR, BR and KL, AE and AC, AC and BL, FH and GJ, ED and DL, FR and RB

8.3 Constructing Perpendicular Bisectors, page 308

1.b) The distance from C and the distance from D to any point on the perpendicular bisector are the same.

2.b) Any point on the perpendicular bisector is the same distance from E as from F.

4.b) The distances from A and from B to the point on the perpendicular bisector are equal.

5.a) Circles intersect only once, at the midpoint of the line segment.

b) Circles do not intersect.

7. Answers may vary. For example: Ceiling or floor tiles

9.a) Connect the points to form a triangle; draw the perpendicular bisector of each side. The point where the bisectors meet is the centre of the circle through the points.

b) Repeat the construction in part a.

8.4 Constructing Angle Bisectors, page 312

1. Yes

2. Yes

3.a) The two angles formed by the bisector will measure 25°.

b) The two angles formed by the bisector will measure 65°.

4. Methods may vary. For example: Use a Mira; use a plastic right triangle; use paper folding.

5. Answers may vary. For example: A ruler and a compass allow for a more accurate construction.

6.c) Two; Opposite angles have the same bisector.

7. c) i) Yes   **ii)** Yes   **iii)** No

8. Answers may vary. For example: Frame of a kite

9.a) The two angles are equal.

b) The centre of the circle is at the intersection of the folded creases.

c) The folding constructed angle bisectors.

Unit 8 Mid-Unit Review, page 314

2.a) AH and CE and FL and GN, AC and HE, FH and EN

b) EH and FL, AC and CE, CE and EH, AH and HE, AH and AC, GN and EH

3.c) Angle measures should be equal.

4.c) Isosceles triangle; AD = BD; CD bisects ∠ADB

5.c) Angle measures should be equal.

8.5 Graphing on a Coordinate Grid, page 318

1. Each grid square represents 5 units. A(10, 15); B(0, 25); C(5, –10); D(–30, 0); E(0, –25); F(0, 0); G(–5, –5); H(–25, 15); J(20, 0); K(–25, –30).

2.a) B, E, and F

b) D, F, and J

c) B, E, and F; H and K

d) D, F, and J; A and H

e) F and G

f) none

3. Answers may vary. For example: Each grid square represents 5 units. O is the origin.

5. Quadrant 3; Quadrant 1; Quadrants 2 and 4

6.c) 16-sided shape with 4 lines of symmetry that intersect at (0, 2). The vertical line of symmetry coincides with the y-axis.

8.a) 8 cm

b) 11 cm

10. Too many to count. For example: A(0,0), B(4, 0), C(5, 3), D(1, 3)

11.b) N(–15, –10)

12.a) Answers may vary. For example: Each grid square represents 2 units.

b) 442 units^2

13. Answers may vary. For example: C(2, 10) and D(–4, 10); C(2, –2) and D(–4, –2); C(–1, 7) and D(–1, 1)

8.6 Graphing Translations and Reflections, page 322

1.a) Reflection

b) Translation

2.a) 3 units left and 9 units up

b) 2 units left and 3 units down

c) 2 units right and 4 units up

d) 3 units left and 2 units down

e) 6 units left

f) 4 units up

3.a) A and C; C is the image of A after a translation 10 units right and 7 units down.

b) B and C; C is the image of B after a reflection in the x-axis.

4. P(2, 3), Q(–2, 2), R(1, –1), S(–1, –3), T(4, –4)

a) P'(–1, 5), Q'(–5, 4), R'(–2, 1), S'(–4, –1), T'(1, –2); the pentagons have the same orientation.

b) P'(2, –3), Q'(–2, –2), R'(1, 1), S'(–1, 3), T'(4, 4); the pentagons have different orientations.

c) P'(–2, 3), Q'(2, 2), R'(–1, –1), S'(1, –3), T'(–4, –4); the pentagons have different orientations.

5.a) A'(1, –3), B'(3, 2), C'(–2, –5), D'(–1, 4), E'(0, 3), F'(–2, 0); the sign of each *y*-coordinate changes.

b) A'(–1, 3), B'(–3, –2), C'(2, 5), D'(1, –4), E'(0, –3), F'(2, 0); the sign of each *x*-coordinate changes.

c) The coordinates of the image should match the patterns in parts a and b.

6.b) A(1, 3), B(3, –2), C(–2, 5), D(–1, –4), E(0, –3), F(–2, 0); A'(–3, 1), B'(–1, –4), C'(–6, 3), D'(–5, –6), E'(–4, –5), F'(–6, –2); Each *x*-coordinate decreases by 4. Each *y*-coordinate decreases by 2.

c) Use the pattern in part b: add the number of units moved to the right or subtract the number of units moved to the left from the *x*-coordinate. Add the number of units moved up or subtract the number of units moved down from the *y*-coordinate.

7.b) The line segments are horizontal. The *y*-axis is the perpendicular bisector of each line segment.

8.b) A'(6, 10), B'(8, 10), C'(8, 8), D'(10, 8), E'(10, 12)

c) A"(–6, 10), B"(–8, 10), C"(–8, 8), D"(–10, 8), E"(–10, 12)

d) Answers may vary. For example: ABCDE and A"B"C"D"E" are congruent, but have different orientations.

9.e) Translation 12 units right and 6 units down

10. Answers may vary. For example: The shape has a line of symmetry that is parallel to the mirror line.

8.7 Graphing Rotations, page 327

1.a) 90° clockwise about the origin or 270° counterclockwise about the origin

b) 180° about the origin

2. The shape was rotated 90° clockwise about the origin (Image 1), reflected in the *x*-axis (Image 2), translated 5 units right and 5 units down (Image 3).

3.a) D(–2, –1), E(–5, –3), F(–1, –5)

b) D'(–1, 2), E'(–3, 5), F'(–5, 1)

c) D"(–1, 2), E"(–3, 5), F"(–5, 1)

d) Yes. The images in parts b and c are the same.

4.a) A'(–2, –5), B'(3, –4), C'(–4, 1)

b) i) OA = OA'

ii) OB = OB'

iii) OC = OC'

c) i) 180° **ii)** 180° **iii)** 180°
All angles measure 180°.

d) A rotation of –180° about the origin

5.a) A'(5, –2), B'(4, 3), C'(–1, –4)

b) i) OA = OA' **ii)** OB = OB'
iii) OC = OC'

c) i) 90° **ii)** 90° **iii)** 90°
All angles measure 90°.

d) A rotation of 270° about the origin

6.a) A(6, 0), B(6, 2), C(5, 3), D(4, 2), E(2, 2), F(2, 0)

b) A'(0, 2), B'(0, 4), C'(–1, 5), D'(–2, 4), E'(–4, 4), F'(–4, 2)

c) A"(–2, 0), B"(–4, 0), C"(–5, –1), D"(–4, –2), E"(–4, –4), F"(–2, –4)

d) Answers may vary. For example: ABCDEF and A"B"C"D"E"F" are congruent and have the same orientation.

7.c) The images coincide. A rotation of 180° is equivalent to a reflection in one axis followed by a reflection in the other axis.
i) Yes
ii) Yes

8. Answers may vary. For example:

b) Rotation about U: R'(2, –4), S'(–3, –4), T'(–3, –1), U(2, –1)

c) Second rotation about U: R"(5, –1), S"(5, –6), T"(2, –6), U(2, –1)
Third rotation about U: R'"(2, 2), S'"(7, 2), T'"(7, –1), U(2, –1)

d) After each 90° rotation counterclockwise about a vertex, the horizontal sides of rectangle RSTU become vertical and the vertical sides become horizontal.

e) Yes. A 90° rotation clockwise about U

9.a) C'(2, –6), D'(3, 3), E'(5, 7); C'(–6, –2), D'(3, –3), E'(7, –5)

b) P'(6, –2), Q'(–3, –3), R'(–7, –5); P'(6, 2), Q'(–3, 3), R'(–7, 5)

c) No

Unit 8 Unit Review, page 335

2.c) The height of ΔCDE

5.a) Scales may vary. For example: Each grid square represents 5 units.

b) A: Quadrant 3, B: Quadrant 4, C: Quadrant 1, D: Quadrant 2

c) Parallelogram; Area = 2500 units²

6.a) Quadrant 4

b) Quadrant 3

c) Quadrant 2

d) Quadrant 1

7.a) i) 12 units   **ii)** 11 units
  b) i) 8 units   **ii)** 6 units
8. (–1, 1) and (3, –1)
9.a) PQRS has only one pair of parallel sides.
  b) P'(7, 1), Q'(11, 1), R'(9, 3), S'(7, 3)
  c) P"(7, –1), Q"(11, –1), R"(9, –3), S"(7, –3)
  d) PQRS and P"Q"R"S" are congruent, but have different orientations.
10.b) P'(3, –1), Q'(7, –1), R'(5, –3), S'(3, –3)
  c) P"(7, –1), Q"(11, –1), R"(9, –3), S"(7, –3)
  Yes, the image remains the same when the translation and rotation are reversed.
11.c) All the images are congruent.
  Under the translation and rotation, the images have the same orientation as quadrilateral ABCD. Under the reflection, the orientation of the image is changed.
12.a) A would be in Quadrant 4, B would be on the negative x-axis, between Quadrants 2 and 3, C would be in Quadrant 2.
  b) Reflection
  c) A 90° or 270° (–90°) rotation
13.b) C'(1, 1), D'(–9, 7), E'(1, 7)
  c) C"(–1, 1), D"(–7, –9), E"(–7, 1)
  d) ABC and A"B"C" are congruent; they have the same orientation.

Unit 8 Practice Test, page 337

4.b) A'(–4, –3), B'(2, –3), C'(1, 1), D'(–3, 0)
  c) A'(2, 4), B'(8, 4), C'(7, 8), D'(3, 7)
  d) A translation 4 units right and 4 units up
  e) The image remains the same.

Cumulative Review Units 1–8, page 342

1.a) $4n + 2$
  c) The graph goes up to the right.
  When the Input number increases by 1, the Output number increases by 4.
2.a) $145; $185
  b) $85 + 2s$
  c) $85 + 4s$
  d) $170 + 2s$
3.a) i) $(+4) + (–5) = –1$
    ii) $(+1) + (–7) = –6$
4.a) High: –4°C; low: –13°C
  b) +9°C or –9°C
5.a) About 9
  b) About 3
  c) About 35
  d) About 249
6.a) $28.89   **b)** Yes; Justin spent $3.89 more.
7.a) 75%, 0.75

  b) 28%, 0.28
  c) 90%, 0.9
  d) 4%, 0.04
8. 20 cm; I assume the medium-sized circles touch the large circle and each other.
9.a) About 58 cm
  b) About 182.21 cm
  c) About 182 cm
  d) About 5 rotations
10.a) 8.64 cm^2
  b) 10.125 cm^2
11.a) $\dfrac{8}{10} = \dfrac{4}{5}$
  b) $\dfrac{5}{12}$
  c) $\dfrac{9}{8} = 1\dfrac{1}{8}$
  d) $\dfrac{13}{12} = 1\dfrac{1}{12}$
12.a) About 2 cups more
  b) $\dfrac{43}{24} = 1\dfrac{19}{24}$ cups
13.a) i) $x - 1 = -2$
    ii) $x + 1 = -3$
  b) i) $x = -1$
    ii) $x = -4$
14.a) $9x = 63; x = 7; 7
  b) $x - 27 = 61; x = 88;$ 88 lures
15.a) $171 000
  b) The mean prize is greater than the median: About 179 571
  c) 79 000
16.a) Mean = 34; median = 33.5; mode = 30
  b) i) Mean = 44; median = 43.5; mode = 40
    The mean, median, and mode increase by 10.
    ii) Mean = 68; median = 67; mode = 60
    The mean, median, and mode double.
17.a) Mean $\doteq 308.4$; median = 305; mode = 305
  b) Outlier: 395
  Mean $\doteq 304.3$; median = 305; mode = 305
  The mean decreases. The median and the mode remain the same.
18.a) Mean = $8.30, median $\doteq 7.88; mode = $7.75
  b) Mean
  c) Outliers: $10.00 and $12.50
  Mean $\doteq 7.97; median = $7.75; mode = $7.75
  The mean and the median decrease. The mode remains the same.
19. False

20.a) $\frac{1}{6}$, $0.1\overline{6}$, about 16%

b) $\frac{100}{100}$, 1, 100%

c) 0, 0%

21.a) There are 48 possible outcomes: 1, 1; 1, 2; 1, 3; 1, 4; 1, 5; 1, 6; 2, 1; 2, 2; 2, 3; 2, 4; 2, 5; 2, 6; 3, 1; 3, 2; 3, 3; 3, 4; 3, 5; 3, 6; 4, 1; 4, 2; 4, 3; 4, 4; 4, 5; 4, 6; 5, 1; 5, 2; 5, 3; 5, 4; 5, 5; 5, 6; 6, 1; 6, 2; 6, 3; 6, 4; 6, 5; 6, 6; 7, 1; 7, 2; 7, 3; 7, 4; 7, 5; 7, 6; 8, 1; 8, 2; 8, 3; 8, 4; 8, 5; 8, 6

b) The outcome of rolling an octahedron does not depend on the outcome of rolling a die.

c) $\frac{4}{48} = \frac{1}{12}$, or $0.08\overline{3}$, or about 8.3%

24. Answers may vary. For example: If both coordinates are positive, the point is in Quadrant 1. If the x-coordinate is negative and the y-coordinate is positive, the point is in Quadrant 2. If both coordinates are negative, the point is in Quadrant 3. If the x-coordinate is positive and the y-coordinate is negative, the point is in Quadrant 4.
If the x-coordinate is 0, the point is on the y-axis. If the y-coordinate is 0, the point is on the x-axis.

25.a) Each grid square represents 5 units.

d) H

26.b) C'(–3, 9), D' (1, 9), E' (1, 3)

c) C" (–3, –9), D" (1, –9), E" (1, –3)

d) C'''(9, –3), D''' (9, 1), E''' (3, 1)

Illustrated Glossary

acute angle: an angle measuring less than 90°

acute triangle: a triangle with three acute angles

algebra tiles: a collective term for unit tiles and variable tiles

algebraic expression: a mathematical expression containing a variable: for example, $6x - 4$ is an algebraic expression

angle: formed by two rays from the same endpoint

angle bisector: the line that divides an angle into two equal angles

approximate: a number close to the exact value of an expression; the symbol $\doteq$ means "is approximately equal to"

area: the number of square units needed to cover a region

array: an arrangement in rows and columns

average: a single number that represents a set of numbers (see *mean, median,* and *mode*)

bar graph: a graph that displays data by using horizontal or vertical bars

bar notation: the use of a horizontal bar over a decimal digit to indicate that it repeats; for example, $1.\overline{3}$ means 1.333 333 …

base: the side of a polygon or the face of an object from which the height is measured

bisector: a line that divides a line segment or an angle into two equal parts

capacity: the amount a container can hold

Cartesian Plane: another name for a coordinate grid (see *coordinate grid*)

central angle: the angle between the two radii that form a sector of a circle

certain event: an event with probability 1, or 100%

chance: a description of a probability expressed as a percent

circle graph: a diagram that uses parts of a circle to display data

circumcentre: the point where the perpendicular bisectors of the sides of a triangle intersect (see *circumcircle*)

circumcircle: a circle drawn through all vertices of a triangle and with its centre at the circumcentre of the triangle

circumference: the distance around a circle, also known as the perimeter of the circle

common denominator: a number that is a multiple of each of the given denominators; for example, 12 is a common denominator for the fractions $\frac{1}{3}, \frac{5}{4}, \frac{7}{12}$

common factor: a number that is a factor of each of the given numbers; for example, 3 is a common factor of 15, 9, and 21

composite number: a number with three or more factors; for example, 8 is a composite number because its factors are 1, 2, 4, and 8

concave polygon: has at least one angle greater than 180°

congruent: shapes that match exactly, but do not necessarily have the same orientation

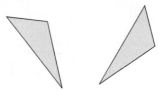

consecutive numbers: integers that come one after the other without any integers missing; for example, 34, 35, 36 are consecutive numbers, so are −2, −1, 0, and 1

constant term: the number in an expression or equation that does not change; for example, in the expression $4x + 3$, 3 is the constant term

convex polygon: has all angles less than 180°

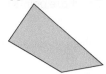

coordinate axes: the horizontal and vertical axes on a grid

coordinate grid: a two-dimensional surface on which a coordinate system has been set up

coordinates: the numbers in an ordered pair that locate a point on the grid (see *ordered pair*)

cube: an object with six congruent square faces

cubic units: units that measure volume

cylinder: an object with two parallel, congruent, circular bases

data: facts or information

database: an organized collection of facts or information, often stored on a computer

denominator: the term below the line in a fraction

diagonal: a line segment that joins two vertices of a shape, but is not a side

diameter: the distance across a circle, measured through its centre

digit: any of the symbols used to write numerals; for example, in the base-ten system the digits are 0, 1, 2, 3, 4, 5, 6, 7, 8, and 9

dimensions: measurements, such as length, width, and height

discount: the amount by which a price is reduced

equation: a mathematical statement that two expressions are equal

equilateral triangle: a triangle with three equal sides

equivalent: having the same value; for example, $\frac{2}{3}$ and $\frac{6}{9}$ are equivalent fractions; 2:3 and 6:9 are equivalent ratios

estimate: a reasoned guess that is close to the actual value, without calculating it exactly

evaluate: to substitute a value for each variable in an expression

even number: a number that has 2 as a factor; for example, 2, 4, 6

event: any set of outcomes of an experiment

experimental probability: the probability of an event calculated from experimental results

expression: a mathematical phrase made up of numbers and/or variables connected by operations

factor: to factor means to write as a product; for example, $20 = 2 \times 2 \times 5$

formula: a rule that is expressed as an equation

fraction: an indicated quotient of two quantities

fraction strips: strips of paper used to model fractions

frequency: the number of times a particular number occurs in a set of data

greatest common factor (GCF): the greatest number that divides into each number in a set; for example, 5 is the greatest common factor of 10 and 15

height: the perpendicular distance from the base of a shape to the opposite side or vertex; the perpendicular distance from the base of an object to the opposite face or vertex

hexagon: a six-sided polygon

horizontal axis: the horizontal number line on a coordinate grid

image: the shape that results from a transformation

impossible event: an event that will never occur; an event with probability 0, or 0%

improper fraction: a fraction with the numerator greater than the denominator; for example, both $\frac{6}{5}$ and $\frac{5}{3}$ are improper fractions

independent events: two events in which the result of one event does not depend on the result of the other event

inspection: solving an equation by finding the value of the variable by using addition, subtraction, multiplication, and division facts

integers: the set of numbers … $-3, -2, -1, 0, +1, +2, +3, …$

intersecting lines: lines that meet or cross; lines that have one point in common

inverse operation: an operation that reverses the result of another operation; for example, subtraction is the inverse of addition, and division is the inverse of multiplication

irrational number: a number that cannot be represented as a terminating or repeating decimal; for example, π

isosceles acute triangle: a triangle with two equal sides and all angles less than 90°

isosceles obtuse triangle: a triangle with two equal sides and one angle greater than 90°

isosceles right triangle: a triangle with two equal sides and a 90° angle

isosceles triangle: a triangle with two equal sides

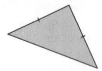

kite: a quadrilateral with two pairs of equal adjacent sides

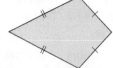

legend: part of a circle graph that shows what category each sector represents

linear relation: a relation whose points lie on a straight line

line graph: a graph that displays data by using points joined by line segments

line segment: the part of a line between two points on the line

line symmetry: a shape has line symmetry when it can be divided into 2 congruent parts, so that one part concides with the other part when the shape is folded at the line of symmetry; for example, line l is the line of symmetry for shape ABCD

lowest common multiple (LCM): the lowest multiple that is the same for two numbers; for example, the lowest common multiple of 12 and 21 is 84

magic square: an array of numbers in which the sum of the numbers in any row, column, or diagonal is always the same

magic sum: the sum of the numbers in a row, column, or diagonal of a magic square

mass: the amount of matter in an object

mean: the sum of a set of numbers divided by the number of numbers in the set

measure of central tendency: a single number that represents a set of numbers (see *mean, median,* and *mode*)

median: the middle number when data are arranged in numerical order; if there is an even number of data, the median is the mean of the two middle numbers

midpoint: the point that divides a line segment into two equal parts

mixed number: a number consisting of a whole number and a fraction; for example, $1\frac{1}{18}$ is a mixed number

mode: the number that occurs most often in a set of numbers

multiple: the product of a given number and a natural number; for example, some multiples of 8 are 8, 16, 24, …

natural numbers: the set of numbers 1, 2, 3, 4, 5, …

negative number: a number less than 0

numerator: the term above the line in a fraction

numerical coefficient: the number by which a variable is multiplied; for example, in the expression $4x + 3$, 4 is the numerical coefficient

obtuse angle: an angle greater than 90° and less than 180°

obtuse triangle: a triangle with one angle greater than 90°

octagon: an eight-sided polygon

odd number: a number that does not have 2 as a factor; for example, 1, 3, 7

operation: a mathematical process or action such as addition, subtraction, multiplication, or division

opposite integers: two integers with a sum of 0; for example, $+3$ and -3 are opposite integers

ordered pair: two numbers in order, for example, (2, 4); on a coordinate grid, the first number is the horizontal coordinate of a point, and the second number is the vertical coordinate of the point

order of operations: the rules that are followed when simplifying or evaluating an expression

origin: the point where the x-axis and the y-axis intersect

outcome: a possible result of an experiment or a possible answer to a survey question

outlier: a number in a set that is significantly different from the other numbers

parallel lines: lines on the same flat surface that do not intersect

parallelogram: a quadrilateral with both pairs of opposite sides parallel

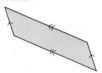

pentagon: a five-sided polygon

percent: the number of parts per 100; the numerator of a fraction with denominator 100

percent circle: a circle divided into 10 congruent sectors, with each sector further divided into 10 parts; each part is 1% of the circle

perimeter: the distance around a closed shape

perpendicular bisector: the line that is perpendicular to a line segment and divides the line segment into two equal parts

perpendicular lines: intersect at 90°

polygon: a closed shape that consists of line segments; for example, triangles and quadrilaterals are polygons

polyhedron (*plural, polyhedra*): an object with faces that are polygons

population: the set of all things or people being considered

positive number: a number greater than 0

prediction: a statement of what you think will happen

prime number: a whole number with exactly two factors, itself and 1; for example, 2, 3, 5, 7, 11, 29, 31, and 43

prism: an object that has two congruent and parallel faces (the *bases*), and other faces that are parallelograms

probability: the likelihood of a particular outcome; the number of times a particular outcome occurs, written as a fraction of the total number of outcomes

product: the result when two or more numbers are multiplied

proper fraction: a fraction with the numerator less than the denominator; for example, $\frac{5}{6}$

pyramid: an object that has one face that is a polygon (the *base*), and other faces that are triangles with a common vertex

quadrant: one of four regions into which coordinate axes divide a plane

quadrilateral: a four-sided polygon

quotient: the result when one number is divided by another

radius (*plural, radii*): the distance from the centre of a circle to any point on the circle

range: the difference between the greatest and least numbers in a set of data

ratio: a comparison of two or more quantities with the same unit

rectangle: a quadrilateral that has four right angles

rectangular prism: a prism that has rectangular faces

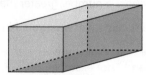

rectangular pyramid: a pyramid with a rectangular base

reflection: a transformation that is illustrated by a shape and its image in a mirror line

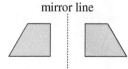
mirror line

reflex angle: an angle between 180° and 360°

regular hexagon: a polygon that has six equal sides and six equal angles

regular octagon: a polygon that has eight equal sides and eight equal angles

regular polygon: a polygon that has all sides equal and all angles equal

related denominators: two fractions where the denominator of one fraction is a factor of the other; their lowest common denominator is the greater of the two denominators

relation: a variable compared to an expression that contains the variable

repeating decimal: a decimal with a repeating pattern in the digits that follow the decimal point; it is written with a bar above the repeating digits; for example, $\frac{1}{11} = 0.\overline{09}$

rhombus: a parallelogram with four equal sides

right angle: a 90° angle

right triangle: a triangle that has one right angle

rotation: a transformation in which a shape is turned about a fixed point

rotational symmetry: a shape that coincides with itself in less than one full turn about its centre is said to have rotational symmetry; for example, a square has rotational symmetry

sample/sampling: a representative portion of a population

sample space: a list of all possible outcomes for an experiment that has independent events

scale: the numbers on the axes of a graph

scalene triangle: a triangle with all sides different

sector: part of a circle between two radii and the included arc

sector angle: see *central angle*

simplest form: a ratio with terms that have no common factors, other than 1; a fraction with numerator and denominator that have no common factors, other than 1

spreadsheet: a computer-generated arrangement of data in rows and columns, where a change in one value results in appropriate calculated changes in the other values

square: a rectangle with four equal sides

square number: the product of a number multiplied by itself; for example, 25 is the square of 5

statistics: the branch of mathematics that deals with the collection, organization, and interpretation of data

straight angle: an angle measuring 180°

surface area: the total area of the surface of an object

symmetrical: possessing symmetry (see *line symmetry* and *rotational symmetry*)

systematic trial: solving an equation by choosing a value for the variable, then checking by substituting

term: (of a fraction) the numerator or the denominator of the fraction

terminating decimal: a decimal with a certain number of digits after the decimal point; for example, $\frac{1}{8} = 0.125$

tetrahedron: an object with four triangular faces; a triangular pyramid

theoretical probability: the number of favourable outcomes written as a fraction of the total number of possible outcomes

three-dimensional: having length, width, and depth or height

transformation: a translation, rotation, or reflection

translation: a transformation that moves a point or a shape in a straight line to another position on the same flat surface

trapezoid: a quadrilateral that has at least one pair of parallel sides

tree diagram: a diagram that resembles the roots or branches of a tree, used to count outcomes

triangle: a three-sided polygon

two-dimensional: having length and width, but no thickness, height, or depth

unit fraction: a fraction that has a numerator of 1

unit price: the price of one item, or the price of a particular mass or volume of an item

unit tile: a tile that represents $+1$ or -1

unrelated denominators: two fractions where the denominators have no common factors; their lowest common denominator is the product of the two denominators

variable: a letter or symbol representing a quantity that can vary

variable tile: a tile that represents a variable

vertex (*plural,* vertices): the corner of a shape or object

vertical axis: the vertical number line on a coordinate grid

volume: the amount of space occupied by an object

whole numbers: the set of numbers 0, 1, 2, 3, …

***x*-axis:** the horizontal number line on a coordinate grid

***y*-axis:** the vertical number line on a coordinate grid

zero pair: two opposite numbers whose sum is equal to zero

Index

A

adding integers, 78
addition equation, 56, 72
algebra, 22 *Math Link*
    solving equations with, 237,
      238, 241–243, 248
algebra tiles, 38–41, 43
    solving equations with,
      231–233, 248
algebraic expressions, 16, 17, 35,
    36, 38–41, 43, 220
angle bisectors, 310–312, 334
area,
    of a circle, 148–150, 167
    of a parallelogram,
      139–141, 144
    of a rectangle, 140, 150
    of a triangle, 143–145, 167
area models, 181
averages, 271–273
    investigating with
      spreadsheets, 276

B

balance-scale models, 226–228,
    238, 242, 248
base, 144, 145, 149, 150
    of a parallelogram, 140, 167
Base Ten Blocks,
    dividing with, 104–106
    multiplying with, 100, 101
benchmarks for comparisons, 91
bisect, 306, 310

C

calculation errors, 155
Carroll diagram, 12
Cartesian plane, 316 *Math Link*
central angle, 161
certain event, 281, 292
chance, 280
circle,
    area of, 148–150, 167
    circumference, 133–135, 167
    diameter, 131

investigating, 130, 131
perimeter, 134
    radius (*pl.* radii), 130
circle graphs, 156–158, 161,
    162, 167
    creating with spreadsheets,
      165, 166
circumcentre, 309
circumcircle, 309
circumference, 133–135, 167
common denominator, 186, 187,
    196, 197, 212
congruent, 321, 326
congruent sectors, 149
congruent shapes, 131
congruent triangles, 144
constant term, 17, 21
coordinate grid, 315–317, 334
coordinates, 315–317
copying errors, 154
Cuisenaire rods,
    modelling mixed numbers,
      204, 205

D

decimals, 120
    adding and subtracting, 96–98
    comparing and ordering,
      91–93
    dividing, 104–106
    from fractions, 86–88
    multiplying, 100, 101
    order of operations with, 108
    relating to fractions and
      percents, 111, 112
denominators,
    related and unrelated, 184
Descartes, René, 316 *Math Link*
diameter, 131, 167
digital roots, 174 *Investigation*
dividend, 105
divisibility, 6–8, 10–12
divisibility rules, 6–8, 10–12, 43
division,
    patterns in, 6–8, 10–12

division sentence, 104
divisor, 105
double prime symbol ("), 324

E

equations, 38–41, 220–223,
    240–243, 248
    preserving equality of,
      229, 248
    reading and writing, 35,
      36, 43
    solving with algebra, 237,
      238, 241–243, 248
    solving with integers,
      231–234
    solving with models,
      226–228
equivalent fractions, 182, 186,
    187, 192, 212
    ordering fractions with, 92
errors, 154, 155
evaluate, 17, 43
experimental probability,
    284, 286

F

factors, 8, 11
fraction circles, 179, 199, 200
fraction strips, 181–183
    adding fractions with, 188
    modelling mixed numbers,
      201
    subtracting fractions with,
      192, 195, 196
fractions, 120
    adding and subtracting, 212
    adding with models, 178,
      179, 181–183
    adding with symbols,
      186–188
    comparing and ordering,
      91–93
    converting to decimals,
      86–88

expressing probability, 280, 281

from circle graphs, 157, 158

relating to decimals and percents, 111, 112

subtracting with models, 191, 192

subtracting with symbols, 195–197

Frayer Model, 290

front-end estimation, 97, 98, 101

G

Games

All the Sticks, 289

Equation Baseball, 245

Packing Circles, 153

graphs,

circle, 156–158

coordinate grids, 315–317

showing relations with, 30–32

H

height, 144, 145, 150

of a parallelogram, 140, 167

hexagon, 178, 192

homework log, 76, 77

I

impossible event, 281, 292

improper fraction, 188, 197, 200, 201, 206, 212

independent events, 285, 292

input/output machine, 25, 26

inspection, 221–223, 234, 248

instructions, 118, 119

integers,

adding, 78

adding on a number line, 60–62, 78

adding with tiles, 56, 57, 78

negative, 52, 53, 56, 57, 60–62, 66–68, 72, 73, 78

opposite, 60, 72, 73, 78

positive, 52, 53, 56, 57, 60–62, 66–68, 72, 73, 78

representing, 52, 53

solving equations with, 231–234

subtracting, 78

subtracting on a number line, 71–73, 78

subtracting with tiles, 66–68, 78

intersecting lines, 303

irrational number, 134, 167

K

key words, 247

L

legend of a graph, 157

line segments, 301, 303–306, 311, 334

linear relations, 31, 32, 43

M

Math Link

Agriculture: Crop Circles, 152

Art, 334

History, 22, 316

Music, 185

Science, 225

Sports, 54

Your World, 120, 288

mean, 258–260, 264, 268, 271–273, 292

median, 263, 264, 268, 271–273, 292

Mira, 304, 305, 312

mixed numbers, 91, 188, 212

adding with, 199–201

subtracting with, 204–206

modelling with Cuisenaire rods, 204, 205

mode, 259, 260, 264, 268, 271–273, 292

models,

adding fractions with, 178, 179, 181–183

subtracting fractions with, 191, 192

multiples, 7, 8

multiplication fact, 174

Investigation

N

negative integer, 52, 56, 57, 60–62, 66–68, 72, 73, 78

notation errors, 155

number lines,

adding fractions on, 182, 183, 188

adding integers on, 60–62, 78

modelling mixed numbers, 201

ordering fractions on, 92, 93

relating decimals, fractions, and percents, 111, 112

subtracting fractions on, 192, 195, 196

subtracting integers on, 71–73, 78

numerical coefficient, 17

O

obtuse triangle, 144

opposite integers, 60, 72, 73, 78

order of operations, 26, 120

with decimals, 108

ordered pair (*see also* coordinates), 316

origin, 316

outcome, 284–286

outliers, 267, 268, 292

P

parallel lines, 300, 301, 334

parallelogram,

area of, 139–141, 144

Pattern Blocks,

subtracting fractions with, 191, 192

pattern rule, 20

patterns,

in decimals and fractions, 87

in division, 6–8, 10–12

in tables, 25–27

relationships in, 20–22

percent circles, 161, 162

percents, 120

expressing probability, 280, 281

from circle graphs, 156–158

relating to fractions and
decimals, 111, 112
solving problems in, 114, 115
perimeter,
of a circle, 134
perpendicular bisectors,
306–308, 322, 334
perpendicular lines, 303,
304, 334
pi (π), 134, 135, 167
pie charts, 165
place-value charts,
ordering decimals on, 93
positive integers, 52, 56, 57,
60–62, 66–68, 72, 73, 78
pressure, 225 *Math Link*
prime number, 90
prime symbol ('), 321
probability, 279–281
experimental, 284, 286
theoretical, 282, 286
protractor, 301, 304

Q
quadrants, 316
quotient, 105

R
radius (*pl.* radii), 130, 167
range, 263, 264, 292
ratio,
expressing probability,
280, 281
rectangle,
area of, 140, 150
reflection, 320–322, 334
computing, 331
reflex angle, 312
related denominators, 184
relations, 21, 25–27, 43
graphing, 30–32
linear, 31, 32, 43

relationships,
in patterns, 20–22
in tables, 25–27
repeating decimals, 87, 88
rhombus, 178, 192, 306, 307, 311
rotations, 325–327, 334
computing, 331

S
sample space, 286
sector angles, 161
sector,
of a circle graph, 156
signatures, 2
simplest form, 182, 187
solutions,
verifying, 227, 228, 233,
238, 241
writing, 210, 211
spreadsheet software, 165, 166
investigating averages
with, 276
study cards, 332, 333
subtracting integers, 78
subtraction, 71
subtraction equations, 67, 72
symbols,
adding fractions with,
186–188
subtracting fractions with,
195–197
systematic trial, 221–223, 248

T
tables,
patterns and relationships in,
25–27
terminating decimals, 87, 88
theoretical probability, 284, 286
thinking log, 14
tiles,
adding integers with, 56,
57, 78

subtracting integers with,
66–68, 78
time zones, 82 *Unit Problem*
transformations, 320, 334
translations, 320–322, 334
computing, 330
trapezoid, 178, 192
tree diagrams, 284–286
triangle, 178, 192
area of, 143–145, 167

U
unit fraction, 189
unit tile, 38
unlike denominators, 212
unrelated denominators, 184

V
variable tile, 38
variables, 16, 17, 21, 36, 43, 233
isolating, 39
Venn diagram, 8, 10
verifying solutions, 227, 228,
233, 238, 241

W
word problems, 246, 247
World of Work
Advertising Sales
Representative, 209
Sports Trainer, 117
writing solutions, 210, 211

X
x-axis, 316, 322, 326, 327, 334

Y
y-axis, 316, 326, 334

Z
zero pairs, 52, 53, 56, 57, 66–68,
231–233

Acknowledgments

The publisher wishes to thank the following sources for photographs, illustrations, and other materials used in this book. Care has been taken to determine and locate ownership of copyright material in this text. We will gladly receive information enabling us to rectify any errors or omissions in credits.

Photography

Cover: Gail Shumway/Getty Images
pp. 2-3 Ian Crysler; pp. 4-5 (left) Canadian Press/Peterborough Examiner/Clifford Skarstedt; (right top) Canadian Press/Calgary Herald/Dean Bicknell; (centre) Ray Boudreau; (bottom) David Young-Wolff; p. 6 Ian Crysler; p. 13 Jan Stromme/Photonica/Getty Images; p. 19 Canadian Press/Carl Patzel; p. 20 Ian Crysler; p. 21 © Dinodia; p. 22 Ludovic Malsant/CORBIS/MAGMA; p. 24 Photodisc/Getty Images; p. 34 Emmanuel Faure/Taxi/Getty Images; p. 39 Dana Hursey/Masterfile www.masterfile.com; p. 46 B & Y Photography/Alamy; p. 48 (top) James Schaffer/PhotoEdit Inc.; p. 48 (bottom) Ian Crysler; p. 49 Tony Freeman/PhotoEdit Inc.; p. 50 (top) Bill Tice/MaXx Images; p. 50 (bottom) Tom Kitchin/Firstlight.ca; p. 51 Tom Bean/CORBIS/MAGMA; p. 52 (top left) Fritz Poleking/footstock/MaXx Images; p. 52 (top right) Mike Copeland/Gallo Images/Getty Images; p. 52 (bottom) Ian Crysler; p. 54 Ryan McVey/Photodisc/Getty Images; p. 56 Ian Crysler; p. 60 Ian Crysler; p. 66 Ian Crysler; p. 72 Ray Boudreau; p. 74 (top) Roy Ooms/Masterfile www.masterfile.com; p. 74 (bottom) Photographer, Gary Herbert; p. 76 Ian Crysler; p. 82 Jamie Squire/Getty Images; p. 83 Corel Collection *China and Tibet*; pp. 84-85 royalty free; p. 96 (top) David Young-Wolff/PhotoEdit, Inc.; p. 96 (bottom) Canadian Press/Dreamworks/Courtesy of Everett Strevt; p. 97 Stone Skyold/PhotoEdit, Inc.; p. 98 Jeremy Woodhouse/Photodisc/Getty Images; p. 100 Ian Crysler; p. 103 Gary Retherford/Photo Researcher's, Inc.; p. 104 Ian Crysler; p. 117 Canadian Press/Aaron Harris; p. 125 Ian Crysler; p. 128 (top) Arthur S. Aubry/Photodisc Collection/Getty Images; p. 128 (bottom) JupiterMedia/Alamy; p. 129 (top left) Hemara Technologies/JupiterImages.com; p. 129 (top right) Galen Rowell/CORBIS; p. 129 (centre) Zedcor Wholly Owned/JupiterImages.com; p. 129 (bottom) Photodisc Collection/Getty Images; p. 130 Ian Crysler; p. 132 LessLIE, Coast Salish artist; p. 133 Ian Crysler; p. 138 Darko Zeljkovic/Canadian Press BLV; p. 143 Ian Crysler; p. 148 Ian Crysler; p. 151 Canadian Press/Jacques Boissinot; p. 152 Jeff Stokoe/Canadian Press RD; p. 154 Ian Crysler; p. 155 Ian Crysler; p. 159 QT Luong/terragalleria.com;

p. 163 Corbis Royalty-Free/MAGMA; p. 164 Hans Blohm/Masterfile www.masterfile.com; p. 165 Bryan and Cherry Alexander/Arctic Photo/Firstlight.com; p. 166 Dale Wilson/Masterfile www.masterfile.com; p. 171 ST-images/Alamy; p. 172 Jeff Greenberg/ PhotoEdit, Inc.; p. 173 Ian Crysler; p. 175 Ray Boudreau; p. 180 Andrew Twort/Alamy; p. 191 Ian Crysler; p. 203 Ray Boudreau; p. 209 Photodisc/Getty Images; p. 211 Ian Crysler; p. 216 Ian Crysler; pp. 218-219 (top) Allana Wesley White/CORBIS; (centre) Stocksearch/Alamy; (bottom right) Jim Craigmyle/CORBIS; p. 220 Michael Newman/PhotoEdit, Inc.; p. 224 David Young-Wolff/PhotoEdit, Inc.; p. 225 Ian Crysler; p. 241 Image100/JupiterImages.com; p. 245 Ian Crysler; p. 247 Ian Crysler; p. 252 Kevin Dodge/CORBIS; p. 253 Ray Boudreau; p. 256 (top) © Comstock Images www.comstock.com; p. 256 (centre) Tim Hall/Photodisc/Getty Images; p. 256 (bottom) David Young-Wolff; p. 256 (background) Dorling Kindersley Media Library; p. 257 (top) Photodisc/Getty Images; p. 257 (bottom) Michael Newman/PhotoEdit, Inc.; p. 257 (background) Johnathan A. Nourok; p. 258 Royalty-Free/CORBIS; p. 261 Noah Graham/Getty Images; p. 262 Ian Crysler; p. 266 John Ulan/Canadian Press CP; p. 267 Ian Crysler; p. 284 Ray Boudreau; p. 289 adapted version of "All the Sticks" game used by permission of Karen Arnason, University of Regina; p. 296 Michael Newman/PhotoEdit, Inc.; p. 297 Ian Crysler; p. 298 Gunter Marx Photography/CORBIS; pp. 298-299 Ann Johansson/CORBIS; p. 299 (left) Don Denton/Canadian Press STRDD; p. 299 (right) age footstock/MaXx Images; p. 300 (top left) Ian Crysler; p. 300 (top centre) Creatas Images/JupiterImages.com; p. 300 (top right) Grant Faint/ Photographer's Choice/Getty Images; p. 300 (bottom) Ian Crysler; p. 301 Ian Crysler; p. 303 (top left) Amy Eckert/Stone/Getty Images; p. 303 (top right) Photos.com/JupiterImages Unlimited; p. 303 (bottom) Ian Crysler; p. 304 Ian Crysler; p. 306 Ray Boudreau; p. 307 Ian Crysler; p. 310 Ian Crysler; p. 311 Ian Crysler; p. 312 Ray Boudreau; p. 320 Ian Crysler; p. 325 Ian Crysler; p. 330 Bryan and Cherry Alexander/Arctic Photo/Firstlight.com; p. 331 Ian Crysler; p. 332 Ian Crysler; p. 334 Digital Vision/Getty Images; p. 339 Ian Crysler; p. 341 Ray Boudreau

Illustrations

Steve Attoe, Pierre Bethiaume, Philippe Germain, Stephen MacEachern, Dave Mazierski, Paul McCusker, Allan Moon, NSV Productions/Neil Stewart, Dusan Petricic, Pronk&Associates, Michel Rabagliati, Craig Terlson, Joe Weissmann, Carl Wiens

The Geometer's Sketchpad, Key Curriculum Press, 1150 65th St., Emeryville, CA 94608, 1-800-995-MATH, www.keypress.com/sketchpad